CHEMISTRY with CHARISMA

24 Lessons That Capture & Keep Attention in the Classroom

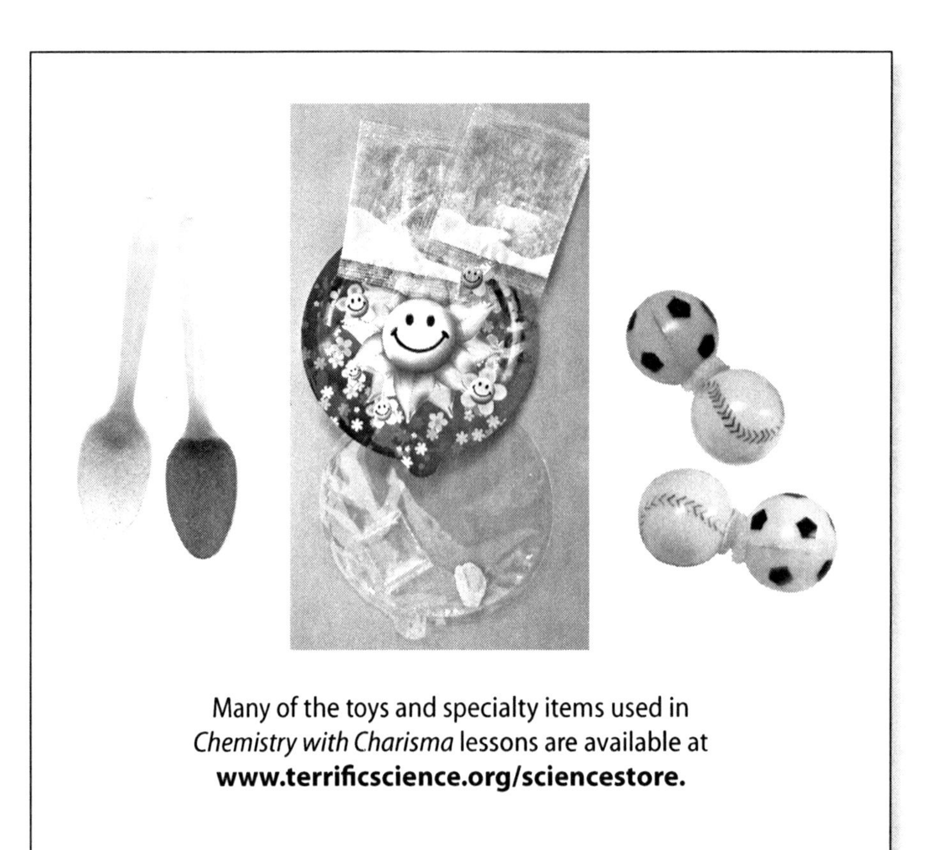

Many of the toys and specialty items used in *Chemistry with Charisma* lessons are available at **www.terrificscience.org/sciencestore.**

This edition of *Chemistry with Charisma* offers limited-time access to downloadable files for student investigations. Files are offered in PDF format for ease of printing as well as in an editable format for customization.

Go to **www.terrificscience.org/charisma** for download instructions.

CHEMISTRY with CHARISMA

24 Lessons That Capture & Keep Attention in the Classroom

Authors

Mickey Sarquis, Director, Center for Chemistry Education & Terrific Science Programs, Miami University, Middletown, OH

Lynn Hogue, Center for Chemistry Education, Miami University, Middletown, OH

Susan Hershberger, Center for Chemistry Education, Miami University, Middletown, OH

Jerry Sarquis, Department of Chemistry & Biochemistry, Miami University, Oxford, OH

John Williams, Department of Chemistry & Biochemistry, Miami University, Hamilton, OH

Terrific Science Press
Miami University Middletown
Middletown, Ohio

Terrific Science Press
Miami University Middletown
4200 East University Boulevard
Middletown, OH 45042

513/727-3269
cce@muohio.edu
www.terrificscience.org

Printed in the United States of America

10 9 8 7 6 5 4 3 2 1

This monograph is intended for use by teachers and properly supervised students. The safety reminders associated with experiments and activities in this publication have been compiled from sources believed to be reliable and to represent the best opinions on the subject as of the date of publication. No warranty, guarantee, or representation is made by the authors or by Terrific Science Press as to the correctness or sufficiency of any information herein. Neither the authors nor the publisher assume any responsibility or liability for the use of the information herein, nor can it be assumed that all necessary warnings and precautionary measures are contained in this publication. Other or additional information or measures may be required or desirable because of particular or exceptional conditions or circumstances.

ISBN: 978-1-883822-55-2

This material is based upon work supported by the National Science Foundation (NSF) under grant numbers DUE-9354378, ESI-9355523, and TPE-9055448, by the Ohio Board of Regents (OBOR) under grant numbers 02-30, 03-27, 04-24, 05-20, 06-25, and 07-26, and by a Science Education Partnership Award from the National Center for Research Resources (NCRR), National Institutes of Health (NIH), under grant number R25RR16301. Any opinions, findings, and conclusions or recommendations expressed in this material are those of the authors and do not necessarily reflect the views of the NSF, OBOR, or NCRR-NIH.

Contents

Acknowledgments

The authors wish to thank the thousands of teachers and their students who have given us feedback and tried out ideas as these activities were developed through 25 years of Terrific Science programming. We also thank the following individuals who contributed to the development of *Chemistry with Charisma.*

Terrific Science Press Design and Production Team

Document Production Managers: Amy Stander, Susan Gertz

Technical Writing: Amy Stander, Susan Gertz, Kate Miller

Technical Editing: Amy Stander, Amy Hudepohl, Joyce Feltz

Production: Anita Winkler, Tom Schaffner, Heather Johnson, Dot Lyon

Photography: Susan Gertz

Cover Design and Layout: Susan Gertz

Introduction

Education is not the filling of a pail, but the lighting of a fire.
William Butler Yeats

As science educators, we all hope to inspire students' motivation and enthusiasm for learning chemistry. We aspire to capture and keep their attention with lessons that provide positive energy, exuberance, and even that certain "magnetic" quality. In short, we want our students to experience "chemistry with charisma." In 1849, a physician named John Scoffern expressed a similar sentiment in *Chemistry no Mystery:*

> *My dear young friends, If I were to present myself before you with an offer to teach you some new game—if I were to tell you an improved plan of throwing a ball, of flying a kite, or of playing leapfrog, oh, with what attention you would listen to me!*
>
> *Well, I am going to teach you many new games. I intend to instruct you in a science full of interest, wonder and beauty; a science that will afford you amusement in your youth, and riches in your more mature years. In short, I am going to teach you the science of chemistry.*

With this book, the authors hope to share a love for teaching chemistry by providing pedagogically rich lessons that are:

- tied to the students' world and meaningful to them;
- motivating and engaging;
- non-biased toward gender or culture; and
- multi-sensory and *fun*.

A wide variety of actively engaging experiences empowers students to gradually formulate their own understandings about abstract, complex chemical systems. As teachers, we need to make sure we are not dissociating fun, hands-on play from minds-on challenges. We need to broaden our teaching repository by interweaving diverse instructional methods to target different learning styles and engage different parts of the brain. We also need to support our students' learning by helping them identify misconceptions, by asking higher-level questions, and by providing a safe environment that encourages students to think critically and take risks.

Engaging Students with Humor and Storytelling

Someone once said that a good teacher is one-third heart, one-third head, and one-third ham. These traits certainly help teachers create connections to chemistry that are relevant to students' lives and popular culture, especially doing so though humor and storytelling.

The following quote from the popular Tom Robbins novel *Even Cowgirls Get the Blues* always gets a laugh and is a great start to a discussion about the role of water on earth: "Human beings were invented by water as a device for transporting itself from one place to another." Remember, a laugh or a giggle is an explicit sign that the information you are transmitting has been received and processed by the learner, which might make you smile in return.

A simple storytelling method can help address the ever-present student questions (implied or spoken) such as "Why do I have to study this stuff? It's so hard. What good is it?" To answer these questions, try telling a story that involves students' favorite pastimes: cars and loud music. Begin by telling students to imagine a picture of themselves out for a good time playing their favorite music in a shiny red sports car.

Then ask, "Where would you be without chemistry?" and read out the items on the following list, pausing after each to give students time to imagine the picture of themselves "out for a good time" changing as they listen.

- Without chemical reactions, the car wouldn't run and the music would stop.
- Without rubber and leather, the car wouldn't have tires or nice leather seats.
- Without paints and coatings, the car wouldn't have its shiny red color.
- Without metals and fiberglass, we'd have no car.
- Without fabrics, you'd be naked.
- Without the chemicals that make up your body, there would be no you.

So where would we be without chemistry? The students are left to ponder this question. Of course they laugh and giggle throughout the telling of this story, but the point is made and you can revisit it throughout the school year.

You may want to have students embellish the "Where would you be without chemistry?" story by doing research online. Prompt students by asking them to find out how the modern conveniences in their lives depend on chemistry. Introduce them to these websites: American Chemical Society (www.acs.org), American Chemistry Council (www.americanchemistry.com), and the educational site by European Chemical Industry Council (www.chemistryandyou.org). This exercise is an opportunity to introduce or reinforce the need for critical evaluation of information found on the web. Students can continue to build their critical thinking and research skills while conducting the web searches suggested in many of the lessons in this book.

Emphasize that using the web effectively requires more than just knowing how to google. When examining web pages and evaluating sources, help students apply these criteria:

- *Authority*: Who are the authors of the web page, or who is responsible for it? What gives them their authority or expertise to write?
- *Accuracy*: Do you have good reason to believe that the information on the site is accurate? Are the facts documented?
- *Objectivity*: What is the author's point of view? What is the purpose of the site?
- *Currency*: When was the information on the page originally written? Has the site been kept up-to-date?
- *Coverage*: Does this site address the topic you are researching? Is the information basic and cursory or detailed and scholarly? Is the information substantial?
- *Value*: Was the page worth visiting? Does the site offer anything informative, unique, or insightful? Is the site free of careless errors, misspelled words, and poor grammar?

Encourage students to extend similar critical thinking to other sources of information, such as books, newspapers, and magazines. Remind students that not everything they'll see in print is true.

Models and Model-Making

A variety of visualization tools and methods can be used to help students process, comprehend, and demonstrate understanding of abstract concepts and phenomena.

Physical models are perhaps the first visualization tools that come to mind. They can help students visualize abstract concepts while acquiring important model development skills. Such physical models are most useful when students are allowed to touch and manipulate them as part of the learning process, an aspect often forgotten in the upper grades. In the "Visualizing Matter" lesson in this book, students investigate the concepts of pure substances and mixtures by manipulating beads and balls as physical models and writing chemical formulas as representational models.

In 1865, German chemist August Wilhelm von Hofmann was the first to make ball-and-stick molecular models such as the model of methane shown above. Photographed by Henry Rzepa, with permission of the Royal Institution of London, in whose collection the model resides.

When thinking about the word "model" in chemistry teaching, don't get stuck imagining the very limited ball-and-stick models that students often see displayed at the front of a classroom. While these can be useful, models in science and in science teaching are so much more. There are mathematical models (often expressed as mathematical formulas or graphs), chemical formulas and equations, diagrams and pictures, and role-playing simulations—all play an essential role in chemistry. One such role-playing simulation included in this book ("Things That Glow") engages students in representing the subatomic difference between phosphorescent and fluorescent materials.

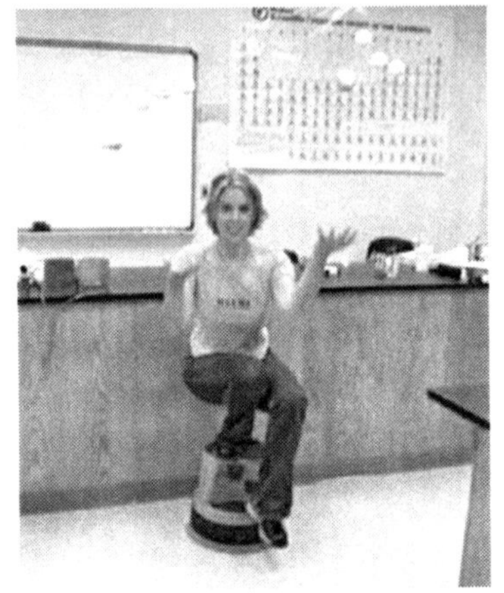

Good models, by definition, have predictive power that enables scientists (and students) to focus attention on the underlying causes of the observed phenomenon and on possible changes in behavior that will result from changes in the system. Importantly, it must be remembered that models are representations and are not reality. By their very nature, models (even very sophisticated ones) have limitations. But awareness and understanding of these limitations add to students' understanding of the real systems the models represent.

Models are essential tools for students as they develop understanding of abstract concepts. While models created by science teachers or textbooks can be useful teaching tools, research suggests that students, especially novice learners, best gain understanding through an active transition from internal mental models gained from interactions with science phenomena to progressively more sophisticated models.

Students should be given the opportunity to express their mental models by creating diagrams, equations, graphs, or physical constructions (ideally made from fun, everyday items). This step allows teachers to see what students are actually thinking (possibly identifying misconceptions) and helps students sharpen their own understanding of the phenomena they have experienced through visual and tactile observations. A technique called storyboarding provides an effective structure for the process of expressing models. After completing a chemistry investigation, have students complete a three-part storyboard in which they first draw the macroscopic phenomena (what they directly observed), then draw their visualization of the phenomena at the particle level, and finally use words to describe how the particle-level representation links to the hands-on phenomena they experienced. The examples below show a storyboard based on the "Syringe Investigations" lesson.

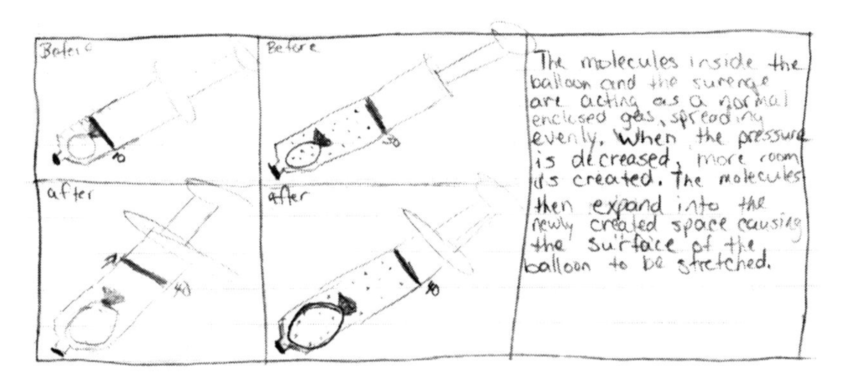

This book offers several other lessons that involve the creative use of models and challenge students to illustrate their mental models of observed phenomena. Through classroom interaction and discussion about the predictive behavior and limitations of student-generated models, teachers can guide students towards understanding of scientifically acceptable models. Our challenge is to help students move through this process.

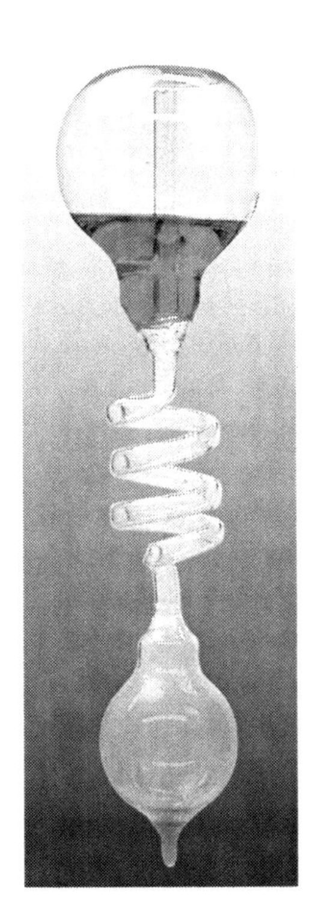

Incorporating Toys and Everyday Materials

While effective inquiry-oriented, hands-on science can take many forms, toys and everyday materials play an important and fun role in the science classroom. In "Teaching Fundamental Aspects of Science Toys" (*School Science and Math Journal,* 1993), O'Brien discusses the pedagogical advantages of using toys to teach science concepts:

- Toys build on and extend a student's out-of-school experiences, intersecting with a student's mental frameworks and enabling him or her to interpret new information in light of pre-existing conceptions.
- Toys are inherently motivational and interactive and simultaneously involve both students and teachers in the fun and mental aspects of science.
- Toys are readily available and generally are lower in cost than conventional science equipment.

This book contains a number of lessons featuring toys and other everyday materials as science tools in order to reunite the fun/hands-on and mental/minds-on aspects of science teaching and provide opportunities to develop process skills, attitudes, and content. For example, in "Investigating a Self-Inflating Balloon," students experiment to determine the chemistry behind this everyday toy. Hand boilers are used both as gas thermometers and intriguing distillation systems. In the lesson "A Cellophane Toy and Its Wrapper," students investigate the chemistry of the materials of a toy cellophane fish and its wrapper and tie their observations to chemical structures.

Developing Understanding with Claims and Evidence

One of the most important aspects of teaching any science is to convey the methodology of scientific investigation in such a way that students develop the skills that are fundamental to scientific inquiry and the scientific way of processing information. The importance of these skills is underscored by their inclusion in the National Science Education Standards and in every state's science standards under headings such as "Science as Inquiry."

While process skills such as observing, sorting, and classifying are important life skills that transcend the discipline of science, science is more than this set of skills; it is a way of looking at, learning about, and interacting with the world. Scientists identify questions to test, systems to experiment on, protocols to follow, and data to collect. This information is then used to formulate claims that can be substantiated by the findings and subsequently shared with the larger community, allowing for open discussion, debate, and scientific argumentation. Scientists must be willing to reflect on this discourse and refine their original claims as new evidence becomes available. The open nature of scientific discourse provides an important safeguard in scientific endeavors.

Students need numerous opportunities to build these skills and experience this process in total. Through experience, the scientific method becomes more than a list they memorize from a textbook, but rather a working system that is an integral part of their lives. Careful observation plays a key role in the process as a catalyst for raising questions and as a means to gather evidence. Students will need to think about what they are observing, discuss their observations with peers, ask questions about what they are seeing, and reflect upon their observations.

Because direct student involvement in this process is so important, we have formulated the lessons in this book to provide opportunities to work through the scientific method. Many lessons prompt students to develop their own testable questions, design experiments, and formulate claims and evidence. To help you, the teacher, the Tips and Instructional Strategies sections include examples of testable questions and experiments that might be undertaken. We expect that you will encourage your students to formulate their own testable questions and experiments and share our suggestions or some of your own when students have trouble with this step. As they progress throughout the year and gain confidence in their scientific investigation abilities, students should be able to formulate their own testable questions, make claims, and provide evidence with increasing ease. Allowing students to ask and strive to answer their own questions gives them a much bigger stake in the outcomes of their investigations, which in turn leads to improved conceptual understanding.

With the time constraints faced in today's classrooms, it could be easy to run out of time before all students present and defend their claims in a larger group. But because this experience is so important to students' growth as scientists, we hope you can work such sharing into your schedule. As an alternative to verbal presentations, students can be asked to write a position statement presenting their claim and their evidence for it. These papers could be peer evaluated for clarity, strength of argument, and other evidence the student-evaluator might be aware of. Other options include students participating in poster sessions; writing informative letters to their families, younger students, or the school board; or developing PowerPoints or YouTube-style videos.

Teachers can maximize student learning by selecting meaningful experiences that grab students' attention, challenge their preexisting beliefs, and encourage the development of testable questions. Discrepant events often provide such opportunities. The surprise of a discrepant event depends on an expectation of the "usual." For example, in the "Where Did the Water Go?" lesson, water is poured into one of three cups, but when all three are turned over, no water pours out. Particularly if students understand the principle of conservation of matter (if you pour water into a cup, it should remain there to be poured out), the absence of water pouring out is surprising. Asking students to pose explanations for why the water didn't pour emphasizes the nature of science. An entertainer might do the same trick and pretend it is magic, but as scientists we are challenged to find logical explanations for events.

Asking students to provide scientific evidence helps them understand the value of using a "control" to provide the direct comparison rather than relying on memory or textbook data. For example, in the "Magic Sand" lesson, students compare and contrast the behavior of water on spoons coated with regular and Magic Sand. While students may assume they already know the behavior of water on regular sand, with both types of spoons in front of them they can make careful, direct observations of both conditions.

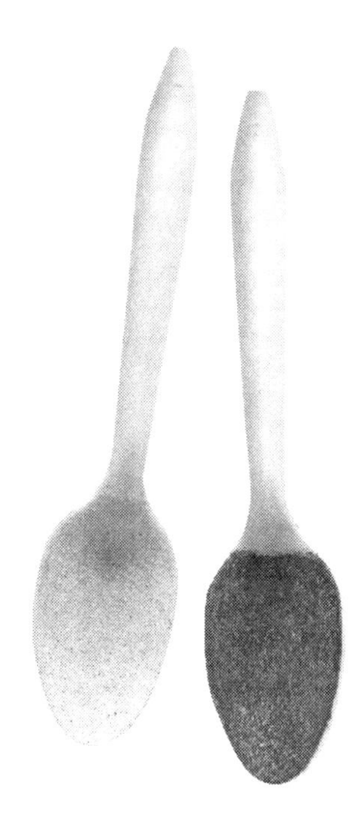

As teachers we must look for opportunities to help students make ties between their observations and important but less obvious aspects of the system being studied. For example, Magic Sand is typically used to teach students about hydrophobic systems. This investigation can also be used as a springboard to lead students to understand that for a substance to be observed as hydrophobic, you not only need the substance, you also need water so the substance can exhibit this effect. This idea is subtle but important pedagogically, because materials do not function in isolation; rather, they function as parts of systems.

Student observations can be greatly enhanced by technology. Digital cameras can record the "before" and "after" for an experiment, especially when a control is not possible, such as in the "Investigations with Hand Boilers" lesson. Digital images also allow viewing at high magnifications. Subtle details can be easily examined, such as the effect of water droplets on colored shaving cream in "Colorful Lather Printing." Recording short movies rather than still images is ideal when an experiment produces fast changes, such as in "Pencil Electrolysis."

Demonstrations versus Hands-On Activities

While hands-on explorations provide students with the most complete experience, in some cases, demonstrations are the prudent alternative. Good reasons for choosing to do demos include

- safety,
- cost and availability of materials,
- limited time, and
- manipulative skills beyond students' abilities.

Because when you do a demo, you are the only one doing it, the challenge before you is to engage your students in verbal discourse that challenges their minds. If possible, we strongly recommend that a part of the lesson also involve a hands-on exploration, as these provide the tactile component otherwise missed. For example, the "Degassing Soda Pop" lesson starts with an attention-grabbing demonstration where a baby bottle nipple inflates dramatically as a result of vigorous shaking of soda pop inside of the baby bottle. The demonstration is used to set the scene for the two subsequent student explorations.

In choosing demonstrations and hands-on experiences for students, keep in mind that individual learners connect with ideas and information in various ways—each student has a unique "intelligence profile." For many educators, the theory of multiple intelligences, beginning with Gardner's *Frames of Mind* (1993), has had a profound impact on thinking and practice. Gardner defines intelligence as the human ability to solve problems or to make something that is valued. He emphasizes that rather than one or two intelligences, all human beings have multiple intelligences, with different strengths in each intelligence area. As a result, each person has a unique intelligence profile. According to Gardner, "It's very important that a teacher take individual differences among kids very seriously.... The bottom line is a deep interest in children and how their minds are different from one another, and in helping them use their minds well." Learners may connect best with

- words (linguistic intelligence),
- numbers or logic (logical-mathematical intelligence),
- pictures (spatial intelligence),
- music (musical intelligence),
- self-reflection (intrapersonal intelligence),
- physical experience (bodily-kinesthetic intelligence),
- social experience (interpersonal intelligence), and/or
- experience in the natural world (naturalist intelligence).

As educators, we are responsible for providing an array of effective learning experiences for our students. The inventive and influential mathematician and cosmologist Sir Hermann Bondi (1919–2005) said, "A scientist is someone whose curiosity survives education's assault on it." By offering our students learning experiences that encourage them in the numerous ways just described, we are helping change the message to "A scientist is someone whose curiosity is nurtured by education's impact on it." But this said, we must remember that science education is not just for future scientists; science is for every student.

It provides them a means of looking at the world and communicating their findings. Today's students are tomorrow's workforce, leaders, and voters; a sound basis in the sciences will support them in all of these endeavors.

References

Armstrong, T. Multiple Intelligences. http://www.thomasarmstrong.com/multiple_intelligences.htm (accessed Mar 2009).

Aubusson, P.; Fogwill, S.; Perkovic, L.; Barr, R. What Happens When Students Do Simulation Role-play in Science? *Res. Sci. Educ.* **1997,** *27* (4) 565–579.

Gardner, H. *Frames of Mind: The Theory of Multiple Intelligences,* 10th ed.; Basic Books: New York, NY, 1993.

Gilbert, J.K., Boulter, C., Eds. *Developing Models in Science Education;* Kluwer Academic: Dordrecht, Netherlands, 2000.

Gilbert, J.K.; Boulter, C.; Rutherford, M. Models in Explanations, Part 1: Horses for Courses? *Int. J. Sci. Educ.* **1998,** *20,* 83–97.

Gobert, J.D.; Buckley, B.C. Introduction to Model-based Teaching and Learning in Science Education. *Int. J. Chem. Educ.* **2000,** *22* (9), 891–894.

Harrison, A.G.; Treagust, D.F. The Particulate Nature of Matter: Challenges in Understanding the Submicroscopic World. In *Chemical Education: Towards Research-Based Practice;* Gilbert, J.K., De Jong, O., Justi, R., Treagust, D., Van Driel, J.H., Eds.; Kluwee Academic: Dordrecht, Netherlands, 2002; pp 189–212.

Henderson, J.R. A Guide to Critical Thinking About What You See on the Web. http://www.ithaca.edu/library/training/think.html (accessed Mar 2009).

MacKinnon, G.; Aucion, J. Refrigeration Dynamics. *Sci. Teacher* **1998,** 65 (9), 33–37.

Norton-Meier, L.; Hand, B.; Hockenberry, L.; Wise, K. *Questions, Claims, and Evidence: The Important Place of Argument in Children's Science Writing;* NSTA Press, Eds.; Heinemann: Portsmouth, NH, 2008.

O'Brien, T. Teaching Fundamental Aspects of Science Toys. *School Sci. and Math. J.* **1993,** *93* (4), 203–207.

Sarquis, A.M. Recommendations for Offering Successful Professional Development Programs for Teachers. *J. Chem. Educ.* **2001,** *78,* 820–823.

Scoffern, J. *Chemistry no Mystery, or, A Lecturer's Bequest;* Harvey and Darton: London, England, 1849.

Smith, M.K. Howard Gardner, Multiple Intelligences and Education. http://www.infed.org/thinkers/gardner.htm (accessed Mar 2009).

Vitz, E. Magic Sand, Modeling the Hydrophobic Effect and Reversed Phase Liquid Chromatography. *J. of Chem. Educ.* **1990,** *67* (6), 512–515.

Safety First

Experiments, demonstrations, and hands-on activities add relevance, fun, and excitement to science education at any level. However, even the simplest activity can become dangerous when appropriate safety precautions are ignored or when the activity is done incorrectly or performed by students without proper supervision. While the activities in this book include cautions, warnings, and safety reminders from sources believed to be reliable, and while the text has been extensively reviewed, it is your responsibility to develop and follow procedures for the safe execution of any activity you choose to do. You are also responsible for the safe handling, use, and disposal of chemicals in accordance with local and state regulations and requirements. The following guidelines will help you and your students to stay safe while exploring the activities in this book.

- Become familiar with the properties of the chemicals and reactions involved in each lesson. To this end, collect and read the Material Safety Data Sheets (MSDS) for all of the chemicals used in your experiments. MSDSs can be obtained upon request from manufacturers and distributors of these chemicals or from a website such as *http://hazard.com/msds/*.
- Read each activity carefully and observe all safety precautions. Determine and follow all local and state regulations and requirements for safety and disposal.
- Always practice activities before using them with your class. This is the only way to become thoroughly familiar with an activity, and familiarity will help prevent potentially hazardous (or merely embarrassing) mishaps. You may also find variations that will make the activity more meaningful to your students.
- You and your students MUST wear appropriate personal protective equipment, including safety goggles, in the laboratory.
- Special safety instructions are not given for everyday classroom materials being used in a typical manner. Use common sense when working with hot, sharp, or breakable objects.
- Caution students never to taste anything made in the laboratory and not to place their fingers in their mouths after handling laboratory chemicals.
- When an activity requires students to smell a substance, instruct them to smell the substance as follows: Hold the container approximately 6 inches from the nose and, using the free hand, gently waft the air above the open container toward the nose. (See figure.) Never smell an unknown substance by placing it directly under the nose.

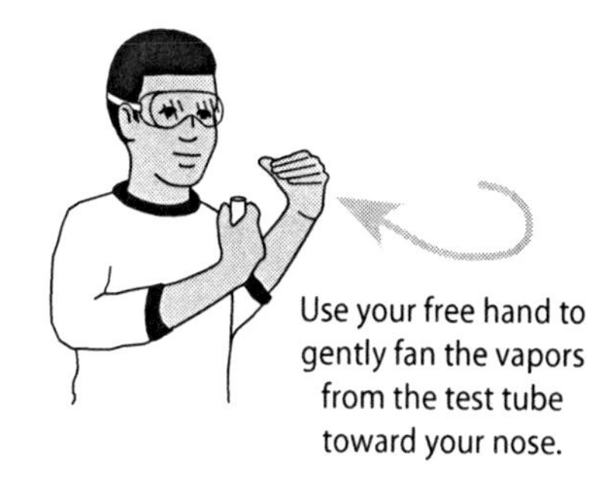

Use your free hand to gently fan the vapors from the test tube toward your nose.

- Remember that you are a role model for your students—your attention to safety will help them develop good safety habits.

SAFETY FIRST

A Cellophane Toy and Its Wrapper

Overview

Students apply the scientific method as they investigate the form and function of the materials used to make this simple toy and its packaging. Part B extends the activity for upper-grade students, as they work to correlate their observations to the chemical structures and FT-IR spectral fingerprints (provided) of these materials.

Key Concepts

- absorption
- evaporation
- experimental design
- hydrophilic substances

National Science Education Standards

Science as Inquiry

Abilities Necessary to Do Scientific Inquiry

- *Students speculate about what causes the movement of the cellophane fish. (5–8)*
- *Students design an experiment to determine which variable has the greatest effect on the behavior of the fish. (5–8, 9–12)*
- *Students develop an explanation based on the results of their investigations. (5–8, 9–12)*
- *In Part B, students compare FT-IR spectral fingerprints for cellulose and polyethylene to understand the correlation of chemical structures with physical properties, spectral data, and commercial applications. (9–12)*

Physical Science

Properties of Objects and Materials

- *The fish is made from cellophane, which attracts water. The wrapper is made of polyethylene, which does not attract water. (5–8)*
- *FT-IR analysis can identify distinguishing characteristics of the materials that comprise the toy fish and its wrapper. (9–12)*

Transfer of Energy

- *Heat energy is transferred from the warmer, body-temperature hand to the cooler, room-temperature fish. (5–8)*

Interactions of Energy and Matter

- *The FT-IR spectra show that different atoms and molecules exhibit peaks at different wavelengths. These peaks can be used to identify a substance. (9–12)*

Science and Technology

Abilities to Do Technological Design

- *Students consider and evaluate the nature of the packaging material used as the wrapper. (5–8)*

Part A: Student Exploration

How does this frisky fish work? Why is it in that wrapper?

Materials

- Fortune Teller Fish in its wrapper
- other materials as needed for the student-designed experiment in step 4, such as
  - source of non-wet, low heat, such as an electric heating pad, hand warmer, or heat pack
  - paper towel
  - water

Procedure

1. Remove the Fortune Teller Fish from its plastic wrapper. Place the fish in the palm of one hand. *What happens? Speculate about what factors might cause this behavior.*
2. Place the fish on a clean, dry table. *What happens? List factors that might cause the difference you observe. Does this change your list of factors from step 1?*
3. Lay the plastic wrapper on your hand. *What happens?* Now put the fish on top of the wrapper on your hand. *What happens?*
4. Experiment to determine which factor has the greatest effect on the behavior of the fish.
5. Now that you've explored the materials used to make this toy fish, imagine that you were the chemist in charge of designing the wrapper for the fish. *What properties or characteristics would be important in making your selection of material?*

Part B: Student Exploration

What's the chemistry of these materials?

Materials

- FT-IR spectrum of the cellulose Fortune Teller Fish (provided)
- FT-IR spectrum of the Fortune Teller Fish wrapper (provided)
- structural formula for cellulose (provided)

Procedure

1. To find out about the chemistry of the materials, you're going to look at the chemical structures of the fish as well as the FT-IR (Fourier Transform Infrared) spectra for the fish and wrapper materials. Don't be intimidated by the term FT-IR—while this is a powerful analytical tool that depicts how atoms in a molecule are connected, it can be simply thought of as providing "fingerprints." You'll be using these fingerprints, which can be used to identify different molecules, to find similarities and differences between the materials used to make the fish and its wrapper. (Note that, while your eyes don't "see" infrared light, different plastic films do respond differently to infrared light, creating different FT-IR fingerprints.)

2. The Fortune Teller Fish is made of cellophane, which is principally cellulose, an organic compound that forms the primary cell wall of green plants. For industrial purposes, cellulose is usually obtained from wood pulp or cotton. Look at the FT-IR of cellulose and find the peaks associated with O-H bonding and C-H bonding. Next look for O-H bonding and C-H bonding in the structural formula for cellulose.

3. *What is the chemical structure for water? Do any parts of the cellulose structure appear to be similar to water's structure?* An important concept in chemistry states that substances that are chemically similar to water are attracted to water. *Based on this concept, suggest a reason for the behavior you observed when the fish was exposed to moisture.*

4. *Predict how the FT-IR spectrum of the wrapper might be different from the spectrum of cellulose.* Now look at the spectrum of the wrapper to check your prediction.

5. Table 1 lists the predominant peaks for common materials used in wrappers. *Which of these materials do you think the Fortune Teller Fish wrapper is made from?*

Table 1: Several Polymer Films and Some Observed Infrared Peaks							
Polymer Film	**Repeating Monomer Unit**	**Wavenumbers (in cm^{-1}) of IR Peaks**					
		3300	**3030**	**2950–2850**	**1700**	**1650**	**1450**
polyethylene	$-CH_2-CH_2-$	No	No	Yes	No	No	Yes
polystyrene	$-CH_2-CH(C_6H_5)-$	No	Yes	Yes	No	Yes	Yes
polyvinyl alcohol	$-CH_2-CH(OH)-$	Yes	No	Yes	No	No	Yes
polyethylene terephthalate	$-CO-C_6H_4-CO-O-CH_2-CH_2-O-$	No	Yes	Yes	Yes	Yes	Yes

Instructor Notes

Tips and Instructional Strategies

- On a very humid day, the fish may move even when placed on the table or on the wrapper. But its movement is rarely as great as when in direct contact with a sweaty hand. If the fish gets really water-logged by a sweaty palm, it may continue to move even on the wrapper or table. Because everyone does not sweat the same way, the behavior of the fish will vary from person to person.
- For Part A, step 4, discuss experimental design as a class. Emphasize the importance of controlling variables. Establish what the variables would be for each testable question proposed, and discuss how students would control these variables. Have students perform their experiments, make claims based on the evidence they collect, and report these claims to the class. Allow for a class discussion and have students reformulate their claims based on this discussion.
- If time does not permit students to design their own experiments in Part A, step 4, you can provide guidance by suggesting that they place the fish on top of a non-wet, mild heat source and record their observations. Ask them if heat affects the fish. Then, similarly test the effect of moisture by laying a damp paper towel (with the excess water squeezed out) on the table and placing the fish on the damp towel. (Take care that students do not overly wet the fish or put the fish directly into water, as this could render it useless.) Then, ask students to draw conclusions as to why the fish behaved the way it did.
- After finishing Part A with the students, you have a good opportunity to initiate a discussion of the scientific method. Ask them what they did in Part A (observe, ask questions, make predictions, identify variables, etc.). Ask them what this process is called (scientific method). This activity also illustrates the non-linearity of the scientific method in a way that students may not have been exposed to previously.
- You may want to encourage students to consider the importance of packaging and the chemists who make decisions about how a product is packaged. Packaging chemists consider factors including the protection of the product, cost, environmental impact, size, consumer appeal, and many others.
- Some cellophane is treated so that it does not absorb water. Thus, attempts to make a homemade version of this toy would require cellophane that has not been treated. (You might want to try a PVA water-soluble laundry bag.)
- Older students continue investigating the Fortune Teller Fish and its wrapper in Part B. We suggest making the FT-IR spectra and the chemical structure of cellulose (located at the end of this activity) into overheads to be shared one by one with students. In step 2, as students look at the spectrum for cellulose, help them to understand that the labeled peaks in the spectra indicate that the molecule has O-H and C-H bonds. When students look at the chemical

structure for cellulose, you may have to remind them that in the shorthand of chemical structures, carbons are present at the five points of the rings even though the Cs are not written. This should help the students to identify areas of O-H and C-H bonding in the structure. In step 3, you may need to help students identify the-OH groups in the cellulose structure as being similar to water.

- Introductory information about FT-IR and its uses is available from the California Institute of Technology (published by the Nicolet Thermo Corporation) at *http://mmrc.caltech.edu/FTIR/FTIRintro.pdf.* (The URL is case-sensitive.)

Explanation

The Fortune Teller Fish curls and twists primarily because it absorbs water produced by the sweat glands in the hand and subsequently loses this water through evaporation. The fish is made of cellophane (mainly cellulose), which is hydrophilic. (*Hydro* means "water" and *philic* means "loving.") As water is absorbed, it moves through small pores in the cellophane and evaporates. The lightness of the cellophane makes the fish very susceptible to air currents, which adds to the "dancing" effect.

In Part A, students explore the Fortune Teller Fish and its packaging. When the fish is on the plastic wrapper on the hand, the fish doesn't move as it does on the bare hand. The wrapper is made of polyethylene, which forms a barrier that prevents the cellophane from absorbing water from the hand. Polyethylene (PE) is a polymer commonly used in consumer goods and packaging. It is made by bonding together thousands of ethylene ($CH_2{=}CH_2$) monomer units.

FT-IR, or Fourier Transform Infrared, is used to identify unknown materials, determine the quality or consistency of a sample, or to determine the amounts of various components in a mixture. FT-IR involves passing infrared radiation through a sample. As the radiation passes through the sample, some is absorbed by the sample. A plot of percent transmittance versus wavenumber will show dips at the wavenumbers where the IR radiation is absorbed. The FT-IR spectrum for a particular molecule is as individual as a human fingerprint—no two unique molecular structures produce the same spectrum (pattern of percent transmittance versus wavenumber).

Answers to Student Questions

Part A

Step 1

a. *The fish moves, curls, twists, and turns. It kind of looks like it's dancing.*
b. *The fish's movement could be caused by the heat, sweat, or saltiness of skin; lotion, perfume, or cologne; light; static electricity; or other factors.*

Step 2

a. *When put on the table, the fish typically flattens out and stops moving.*

b. The movement of the fish must have been caused by something that is present on the palm of the hand that is not present on the table.

c. Referring back to answer 1b, the light can now be eliminated as a cause, as the light was present in both cases. Possible factors still include heat, sweat, or saltiness of sweat; lotion, perfume, or cologne; or static electricity.

Step 3

a. The wrapper lies flat without any visible change or effect.

b. The fish lies on the wrapper without any visible change or effect.

Step 5

The wrapper should be waterproof, low cost, lightweight, attractive, and have minimal environmental impact.

Part B

Step 3

a. The chemical structure of water is a bent molecule, with oxygen in the center and two O-H bonds. The H-O-H bond angle is 109 degrees, which provides the bent shape.

b. Each cellulose structure has three O-H bonds; two are attached to the rings and one is attached to a carbon attached to the ring. These O-H bonds are similar to the O-H bonds in water.

c. The attraction of water to the cellophane fish caused the bottom side of the fish, on either a sweaty hand or a damp paper towel, to absorb a small amount of water, causing the observed curling.

Step 4

The FT-IR for the wrapper will not show the -OH peaks.

Step 5

The wrapper must be made of polyethylene.

Reference

Thermo Scientific Website. http://www.thermo.com/eThermo/CMA/PDFs/Articles/articlesFile_12268.pdf (accessed Mar 2009), Introduction to Fourier Transform Infrared Spectrometry.

FT-IR Spectrum
Cellulose Fortune Teller Fish

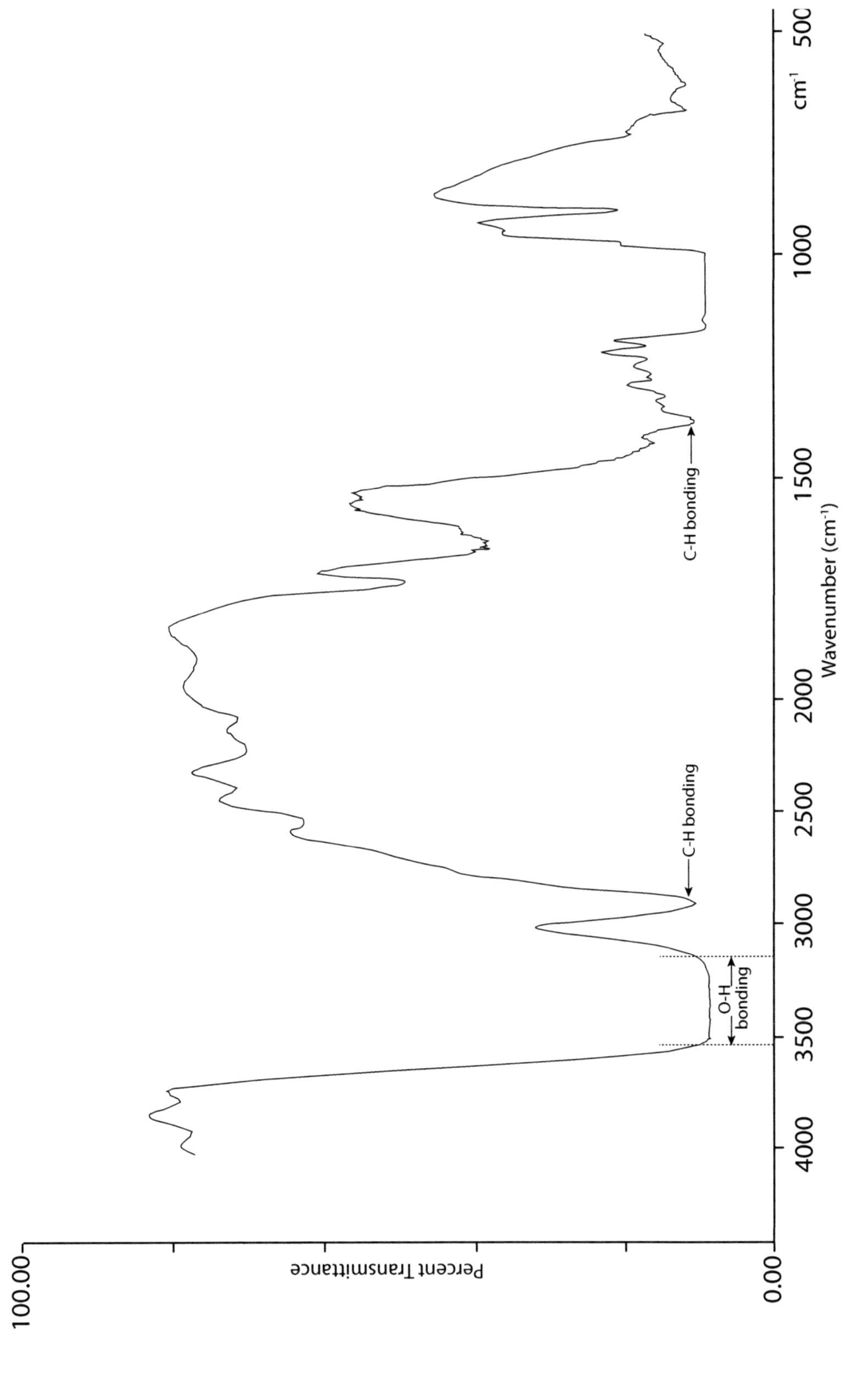

Percent Transmittance

100.00

0.00

4000 3500 3000 2500 2000 1500 1000 500 cm⁻¹

Wavenumber (cm^{-1})

O-H bonding

C-H bonding

C-H bonding

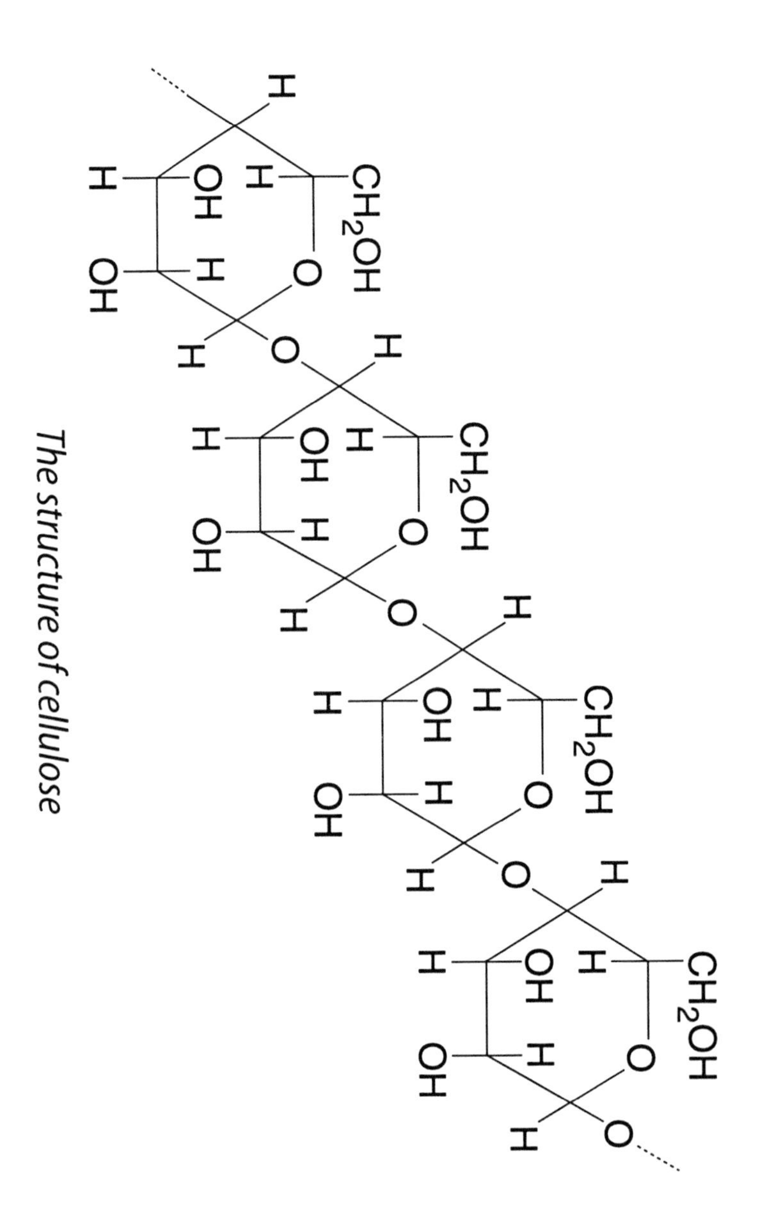

The structure of cellulose

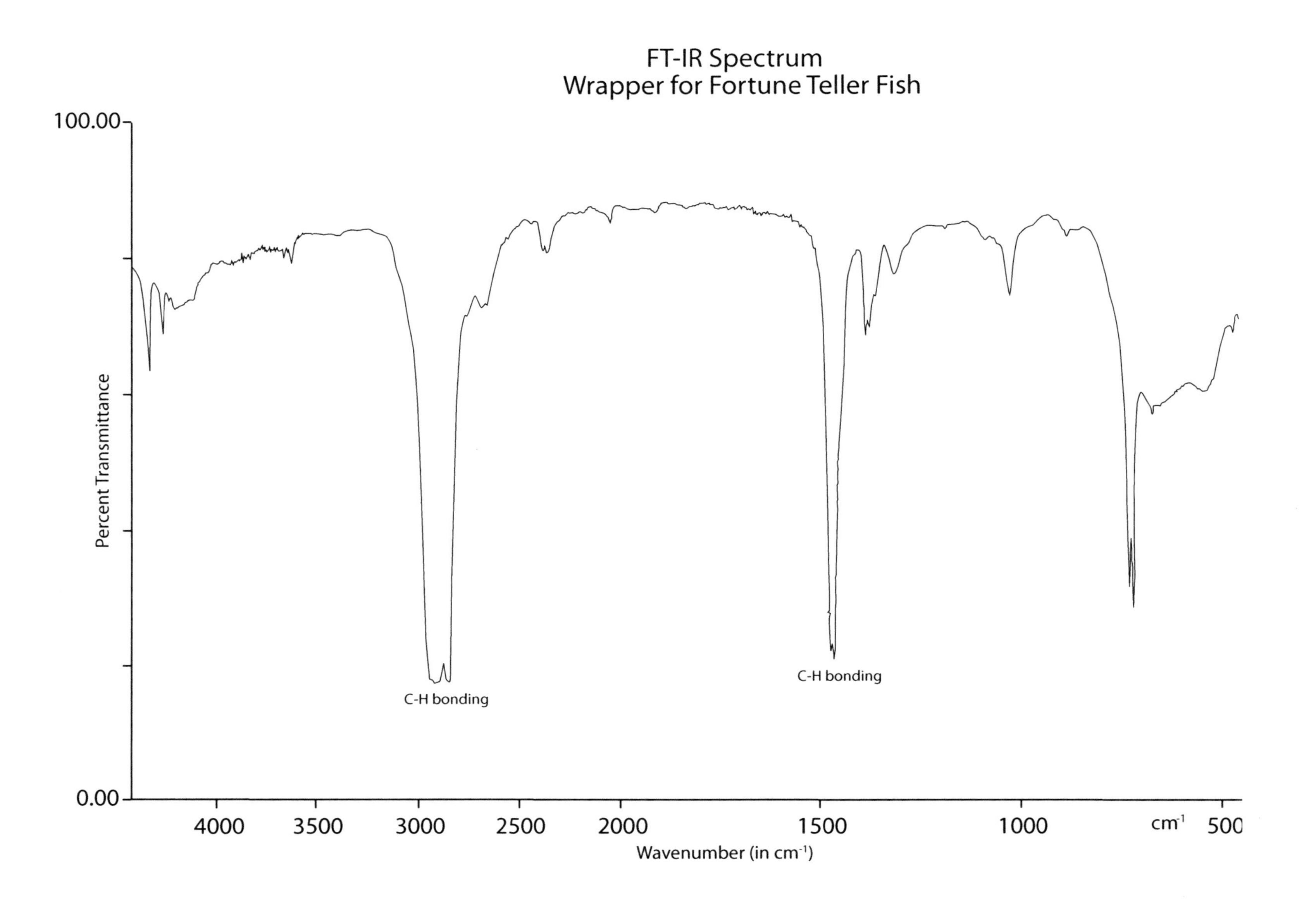
FT-IR Spectrum
Wrapper for Fortune Teller Fish
100.00
Percent Transmittance
0.00
C-H bonding
C-H bonding
4000
3500
3000
2500
2000
1500
1000
cm-1
500
Wavenumber (in cm-1)

Where Did the Water Go?

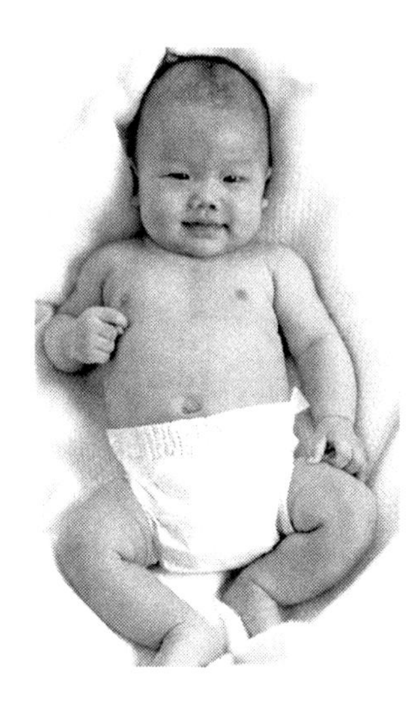

Overview

This lesson begins with a teacher demonstration of a nifty "magic trick" that introduces students to the water-absorbing properties and real-world applications of sodium polyacrylate. The student explorations engage students in determining the amount of water the super-absorbing materials can hold, the effect of different ionic and non-ionic solutes in water on the absorbing capacity of the polymer, and the water-absorbing capacity of the polymer in disposable diapers.

Key Concepts

- absorption
- discrepant events
- experimental design
- hydrophilic substances
- ionic substances
- polymers

National Science Education Standards

Science as Inquiry Standards

Abilities Necessary to Do Scientific Inquiry

- *Students speculate about what causes water to "disappear" after being poured into a cup. (5–8)*
- *Students develop simple hypotheses to explain what causes the water in the cup to "disappear." (5–8)*
- *Students develop explanations based on the results of their observations. (5–8, 9–12)*
- *Students compare different compounds to see which has the greatest effect on de-gelling the super-absorbing polymer. (5–8, 9–12)*
- *Students design an experiment to compare the water-holding capacities of different brands of diapers. (5–8, 9–12)*
- *Students design experiments to determine how much water and how much salt water disposable diapers can absorb. (5–8, 9–12)*

Physical Science

Structure and Properties of Matter

- *Sodium polyacrylate is a hydrophilic polymer capable of absorbing large quantities of water. (9–12)*
- *The polymer absorbs less water when water-soluble ionic compounds are present. (9–12)*

Science and Technology

Abilities to Do Technological Design

- *Students consider and evaluate the nature of super-absorbing polymers. (5–8)*
- *Students learn about two commercial applications of super-absorbing powders (diapers and fire retardants). (5–8)*

Part A: Teacher Demonstration

Is the hand quicker than the eye?

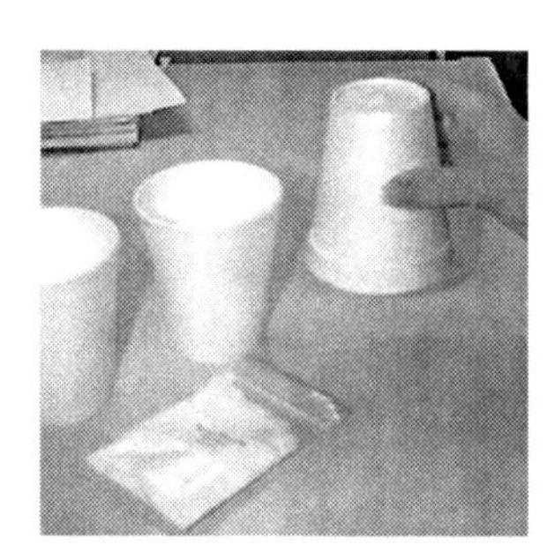

Materials

- 3 identical Styrofoam® cups
- super-absorbing powder (sodium polyacrylate)
- water
- clear plastic cup
- stir stick

Procedure

Avoid ingesting or inhaling the super-absorbing powder. Should eye contact occur, rinse the area with lots of water.

1. In preparation for the demonstration, select one of the three identical Styrofoam cups to be the "reaction cup." (The other two cups function as distracters in this trick.) Sprinkle enough super-absorbing powder into the reaction cup to barely cover the bottom of the cup.

2. Arrange the reaction cup and the two distracter cups in a row on a table. (When you perform the trick, be sure that your class stands or sits a few feet from the table so they can't see into the cups.) Tell your audience that you want to test their powers of observation, so they will need to watch carefully.

3. Pour about 80 mL ($^1/_3$ cup) water into one of the distracter cups (not the reaction cup). Tell your audience that you will mix up the cups and they are to keep their eyes on the one with the water. Slowly move the cups around a few times, exchanging their positions. Ask the class which cup contains the water.

4. Reinforce that they are correct, as you pick up the cup and pour the water into the reaction cup. Take a little time to talk about the importance of good observations and how you are going to move the cups faster next time. (This banter allows time for the water and powder to form a gel.)

5. Shuffle the cups around again and ask the class which cup has the water in it. Turn over one of the empty cups, then the other empty one, and finally, slowly turn over the reaction cup as though you are anticipating an impending spill. Your class will be surprised when no water comes out.

6. Ask the class where the water went. Lead students through a discussion of conservation of matter and how what happened was science, not magic. Let students try to figure out how the trick was done. Ask them to predict what would happen if you placed a stir stick into an empty cup. (*The stick will tilt against the side of the cup.* Do this to show them.) Ask them what

would happen if the stick were put into the cup they thought had the water in it. Do this and reveal that the stick stands straight up in the cup, providing evidence something else besides water is in the cup.

7. Lift the gelled super-absorbing material out of the cup and put it in a zipper-type plastic bag so you can pass it around for students to see up close. Explain that you placed a small amount of super-absorbing polymer in the reaction cup before the demonstration.

8. Repeat the gelling process in a clear plastic cup to show how quickly the gelling takes place. Ask students if they know a practical application for this polymer. (Super-absorbing polymers are used in diapers and as a fire retardant.)

Part B: Student Exploration

How much water will the polymer hold?

Materials

- super-absorbing powder (sodium polyacrylate)
- 300- to 355-mL (10- to 12-ounce) clear plastic cups
- measuring spoons
- disposable diapers that contain super-absorbing polymers
- tap water
- other materials as needed for the student-designed experiment

Procedure

Avoid ingesting or inhaling the super-absorbing powder. Should eye contact occur, rinse the area with lots of water.

1. Experiment to find out how much water the polymer will hold by pouring enough of the super-absorbing powder (sodium polyacrylate) into a cup to barely cover the cup bottom. Spread the polymer evenly on the bottom of the cup and slowly add water 15 mL (1 tablespoon) at a time. Gently stir and then wait about 10 seconds between additions. *Record your observations. How much water were you able to add before the gel lost its form and began flowing?* Compare your findings with others in the class.

2. Design an experiment to determine which brand of disposable diaper holds the most water and compare your results to your findings in step 1.

Part C: Student Exploration

How do salt and other additives affect the gelled polymer?

Materials

- super-absorbing powder (sodium polyacrylate)
- clear plastic cups
- measuring spoons and cups
- sodium chloride (table salt, NaCl)
- sodium bicarbonate (baking soda, $NaHCO_3$)
- sugar
- table salt substitute that contain potassium chloride (KCl)
- water
- disposable diapers that contain super-absorbing polymers
- other materials as needed for the student-designed experiment

Procedure

1. Gel some polymer using the maximum amount of water you discovered the polymer would hold in Part B.
2. Divide the gelled polymer into four or five cups, putting roughly equal amounts in each.
3. Sprinkle a pinch of table salt (NaCl) over the polymer gel in one of the cups and vigorously stir. Observe the results. Add and count successive pinches of salt until adding another pinch of salt causes no further effect. *Record the amount of table salt added.*
4. Repeat step 3 using sugar, baking soda, table salt substitute, or similar substances. Compare your observations to those in step 3. *Which substance took the least amount to de-gel the polymer? Which, if any, of the substances did not de-gel the polymer? Based on your observations, make a claim about the types of substances that work to de-gel the polymer.*
5. Design an experiment to determine how much salt water a disposable diaper will hold. Compare the results to those you obtained using tap water in Part B.

Instructor Notes

Tips and Instructional Strategies

- For Part A, you may want to cover the outside of the Styrofoam cups with contact paper or construction paper to prevent students from seeing a shadow in the reaction cup (an issue in certain lighting conditions).
- For Part A, step 2, you may want to place the cups on a box, which will serve as another distracter. Students may guess that the water went into the box. Then you can open the box to show them the water is not there.
- To add a quantitative dimension to Part B, have students measure (mass or volume) the amount of super-absorbing powder to be used in step 1. Then based on their results in steps 1 and 2, students should be able to estimate the amount of super-absorbing material in the diaper they tested in step 2.
- For Parts B and C, discuss experimental design as a class. Emphasize the importance of controlling variables. Establish what the variables would be for each question proposed, and discuss how students would control these variables. One strategy is to place the diaper over a bucket and then add liquid in 15-mL (1-tablespoon) increments until the diaper no longer absorbs added liquid. For Part C, you may want to coach the students to use different concentrations of NaCl solutions and compare the results.
- Share with students the story of the development of a fire retardant gel that contains sodium polyacrylate. (See the Explanation.) Emphasize how the firefighter acted as a scientist to discover an explanation for an unexpected observation and to apply the knowledge gained in new ways.

Explanation

The powder used in this activity is a super-absorbing copolymer that contains sodium polyacrylate. Sodium polyacrylate is a synthetic hydrophilic polymer. (The chemical structure for the hydrophilic unit of this copolymer is shown in Figure 1.) Hydrophilic means "water loving." In Part A, the powder was placed into one of the cups before the demonstration began. The powder quickly gelled when water was added because the water was attracted to the numerous acrylate groups within the polymer. Sodium polyacrylate can absorb several hundred times its weight in water, depending on the purity of the water used. The swelling that results helps to account for the fact that the gel did not spill out of the cup when it was inverted. In Part B, students explore how much water the polymer can hold and design an experiment to determine which brand of diaper holds the most water.

```
    H   H
    |   |
 ---C---C---
    |   |
 O==C   H
    |
    O⊖Na⊕    n
```

Figure 1: Chemical structure of sodium polyacrylate unit in the copolymer

Sodium chloride, salt substitute that contains potassium chloride, baking soda, and other water-soluble ionic compounds interfere with the polymer's ability to gel, which accounts for the observations in Part C where the gel was turned into a slurry upon addition of these materials. Non-ionic compounds, such as sugar, do not appear to affect the gel in the same manner. While sodium polyacrylate can absorb about 300 times its weight in tap water (and 800 times its weight in distilled water), it can only absorb about 50 times its weight of a 1% aqueous sodium chloride solution. In Part C, students design their own experiments to determine how much salt water a diaper will hold and then compare their results to those obtained with tap water in Part B. Students should find that a diaper holds significantly less salt water.

Super-absorbent materials were first developed for use in agriculture (as vehicles for water retention) and in the oil industry (as additives in drilling fluid in offshore operations.) Today, super-absorbent materials are also used in diapers, personal care products, and as fire retardants.

While the potential of polyacrylate for fire fighting was recognized in the 1960s (when a patent for a water-immobilizing gel was registered), the idea was not widely applied until independently rediscovered in the early 1990s by John Bartlett, an observant firefighter in Florida. While examining the debris of a fire, he discovered that a used disposable diaper had not burned. Based on this observation, Bartlett and a team of firefighters and chemists worked for five years to develop an aqueous slurry of this polymer that could function as a fire-blocking gel. Now widely used, this slurry can be applied to structures, vehicles, fuel tanks, vegetation, and other objects exposed to fire. Because the water in the gel has a high heat capacity, it prevents the substance it is applied to from reaching kindling temperature. The slurry is even dropped from helicopters or airplanes during wildfires to create firebreaks. Fire-blocking gel has been called "a quantum leap in firefighting" by William Kramer, Fire Science Professor at the University of Cincinnati. Students may want to learn more about this fire-fighting gel at *www.firegel.com.*

Answers to Student Questions

Part B

Step 1

The water is absorbed and a gel is formed. Students should find they were able to add about 150 mL (10 tablespoons) of water before the gel lost its form.

Part C

Step 3

Answers may vary, but our testing showed that about three pinches of salt had the maximum effect on the gelled polymer.

Step 4

Table salt (sodium chloride) took the least amount to de-gel the polymer. Sugar had little effect on the gel.

References

Bartlett, J.B. Fire Gels–Breakthrough Technology for Structure Protection in the Wildland/Urban Interface. Presented at the 2nd International Wildland Fire Ecology and Fire Management Congress, Orlando, FL, November 2003; 4A.7.

Criswell, B. Ions or Molecules? Polymer Gels Can Tell. *J. Chem. Educ.* **2006,** *83,* 567A.

Schwarcz, J. *The Genie in the Bottle;* Henry Holt and Company: New York, NY, 2002.

Woodward, L. *Polymers All Around You!,* 2nd ed.; Terrific Science Press: Middletown, OH, 2002.

Syringe Investigations

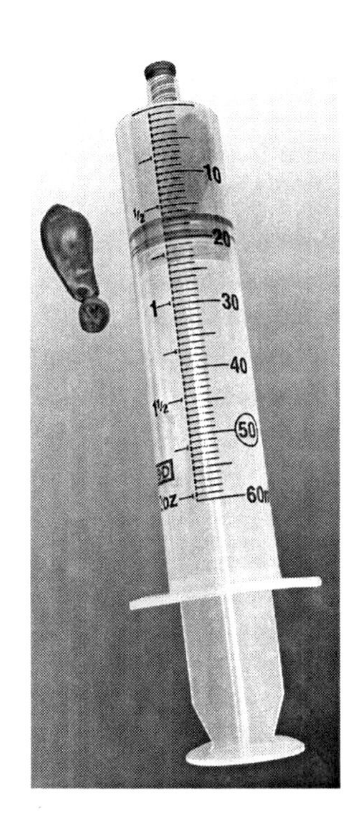

Overview

Students do several hands-on activities that use a common gadget to illustrate the relationships of pressure and volume of gases.

Key Concepts

- air takes up space
- atmospheric pressure
- compressibility
- experimental design
- gases
- gas laws
- particle nature of matter
- pressure
- pressure-volume relationship of gases
- volume

National Science Education Standards

Science as Inquiry

Abilities Necessary to Do Scientific Inquiry

- *Students make observations about the behavior of air and other materials in a plastic syringe and ask questions based on their observations. (5–8)*
- *Students use a syringe to measure quantities of air. (5–8, 9–12)*
- *As an extension, students pose testable questions and design experiments to answer these questions. (5–8, 9–12)*

Understandings about Scientific Inquiry

- *Students ask and answer questions about the behavior of the syringe and compare their answers with scientific knowledge about the pressure-volume relationships of various materials. (9–12)*

Physical Science

Structure and Properties of Matter

- *Students learn that gas pressure results from the collision of gas particles with the walls of their container. (9–12)*
- *Students learn about the particle nature of matter in the gas phase as they explore what happens to pressure as the volume of a gas changes (Boyle's law). (9–12)*

Part A: Student Exploration

Can you feel air pressure?

Materials

- safe plastic syringe with 60-cc capacity and airtight rubber cap

Procedure

1. Look at Figure 1 to identify the parts of the plastic syringe. The numbers on the barrel show the volume of material in the syringe in units of cubic centimeters (cc or cm3) or milliliters (mL). Pull about 30 cc of air into the syringe by pulling the plunger back until the front edge of the gasket lines up with the 30-cc volume marking. Put your finger close to the open tip and push the plunger into the barrel. *What do you feel?*

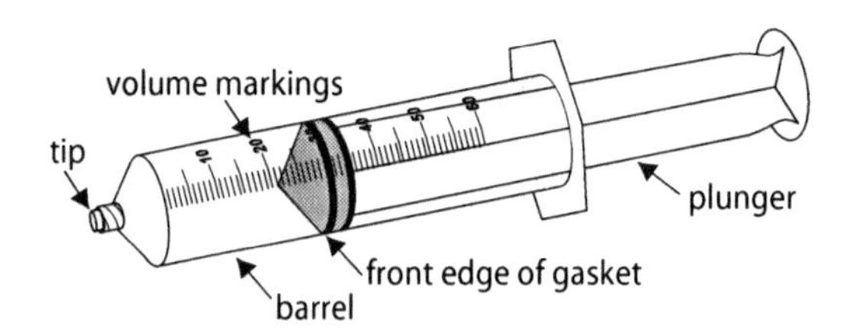

Figure 1

2. Pull about 50 cc of air into the syringe. Cover the syringe tip firmly with your finger and leave it there until the end of this step. Push the plunger in as hard as you can. *What volume can you compress the air to?* While still holding onto the syringe barrel and with your finger still over the tip, release the plunger. *What happens? Why?*

3. Repeat step 2 but do not release the plunger. Instead, remove your finger from the tip once you have compressed the gas. *What do you hear? Is this sound due to air moving into or out of the syringe? Why do you think this?*

4. Pull 10 cc of air into the syringe. Place the airtight cap onto the syringe. Pull the plunger back as far as you can but not all the way out. *What do you feel?* Let go of the plunger. *What happens? Why do you think this happens?* Repeat several times. *Do you always get the same result?*

5. With the rubber cap still on the syringe tip, try to pull the plunger completely out of the barrel. *What do you hear when the plunger comes out of the barrel? Is there air movement in this process? If so, in which direction?*

Part B: Student Exploration

Compare and contrast balloons and marshmallows.

Materials

- safe plastic syringe with 60-cc capacity and airtight rubber cap
- 3 small balloons
- 2 miniature marshmallows
- pen or marker

Procedure

Figure 2

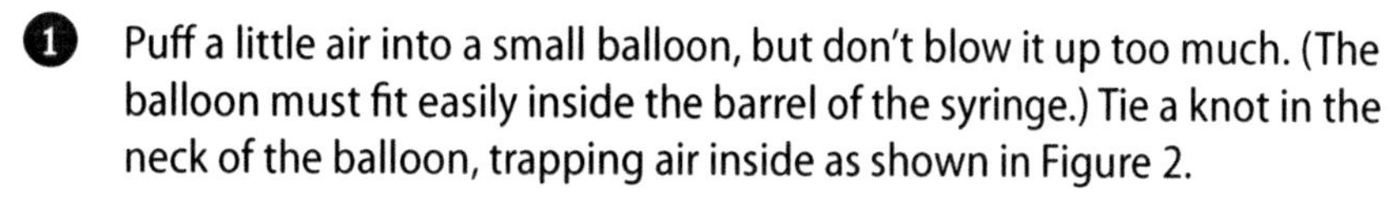

1. Puff a little air into a small balloon, but don't blow it up too much. (The balloon must fit easily inside the barrel of the syringe.) Tie a knot in the neck of the balloon, trapping air inside as shown in Figure 2.

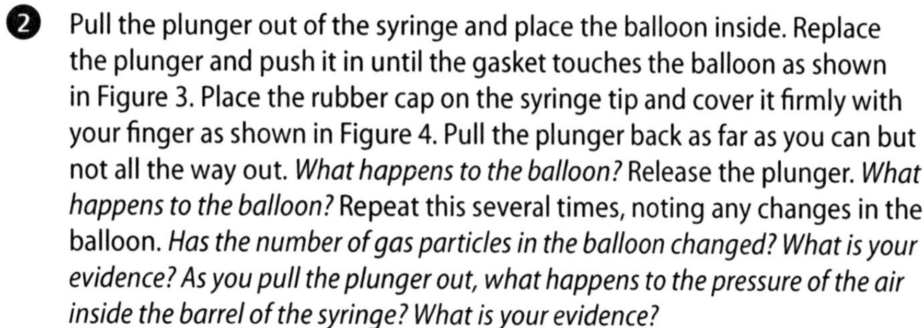

2. Pull the plunger out of the syringe and place the balloon inside. Replace the plunger and push it in until the gasket touches the balloon as shown in Figure 3. Place the rubber cap on the syringe tip and cover it firmly with your finger as shown in Figure 4. Pull the plunger back as far as you can but not all the way out. *What happens to the balloon?* Release the plunger. *What happens to the balloon?* Repeat this several times, noting any changes in the balloon. *Has the number of gas particles in the balloon changed? What is your evidence? As you pull the plunger out, what happens to the pressure of the air inside the barrel of the syringe? What is your evidence?*

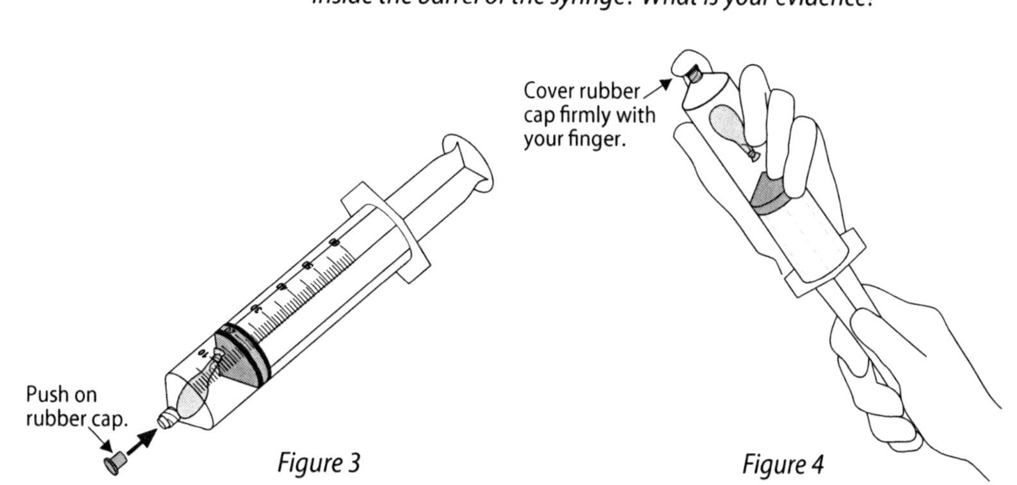

Figure 3 *Figure 4*

3. *What would you have to do to increase the pressure inside the syringe? Predict what would happen to the balloon if you did this.* Try it and see what happens. *Compare your observation to those from Part A.* Take the rubber cap off the syringe, remove the plunger, and discard the balloon.

4. Select two miniature marshmallows about the same size. For future reference, trace around the circumference of each marshmallow on a piece of paper. Use a pen or marker to draw faces on the marshmallows.

5. Put one of the marshmallows inside the syringe. Leave the other marshmallow out to serve as your control. Put the plunger back in the syringe barrel. Push the plunger in until the gasket just touches the marshmallow but doesn't squeeze it. Place the rubber cap on the syringe tip and hold it firmly with your finger.

6. Reduce the pressure inside the syringe by pulling the plunger back (but not all the way out) and holding it in the "out" position. *What happens?* Compare the size of the marshmallow in the syringe with that of the control. Let go of the plunger. *What happens to the marshmallow now?* Repeat the pull and release several times. Remove the marshmallow from the syringe and compare it to the control and to your tracing. *How are they different? Why?*

   How are the behaviors of the marshmallow and the balloon inside the syringe similar? How are they different?

Instructor Notes

Tips and Instructional Strategies

- In Part A, it's crucial that each student have the experience of trying to manipulate the syringe—this tactile, haptic experience strongly supports student learning and retention of Boyle's law and air pressure. You may want to tell students that Robert Boyle was the first to report that the springiness of air was similar to the springiness of metals.
- After Part A, step 2, ask the students to make a series of drawings of what they observed at the macroscopic level. Then ask them to draw what they visualize is happening at the molecular level. Finally ask them to write a short paragraph explaining the relationship between their molecular-level visualization and the macroscopic phenomena. This three-step process allows the student to concentrate on the particle nature of matter and how the particles create the pressure. A sample is provided here.

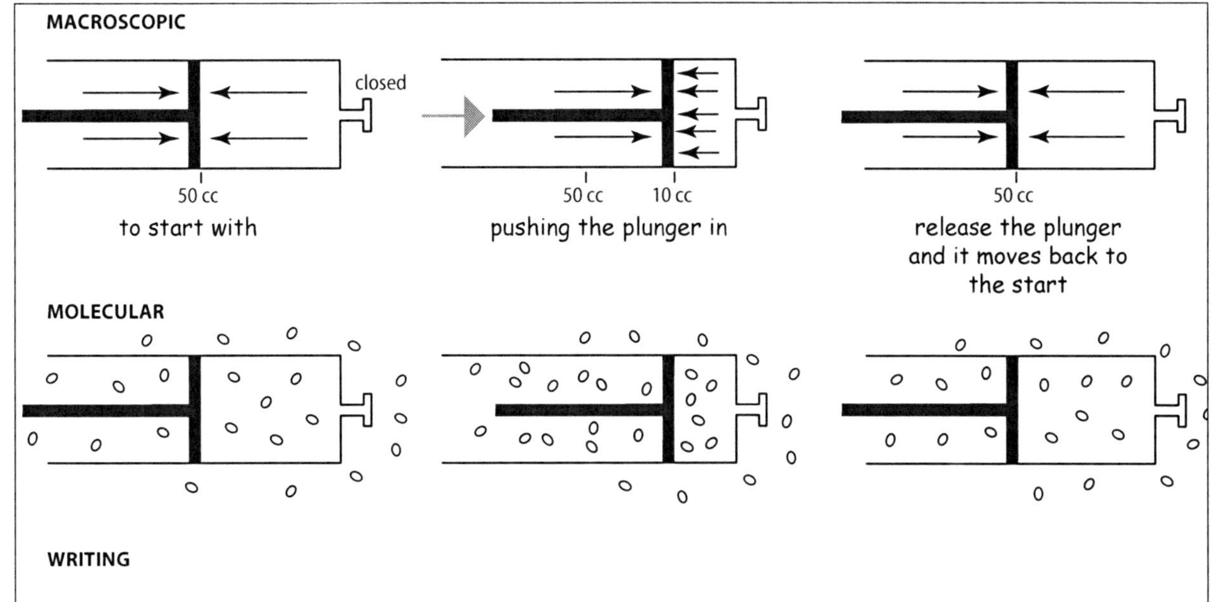

The particles of trapped gas inside the syringe hit the inside walls creating a pressure that is at first equal to atmosperic pressure. When you push the plunger in, the number of trapped particles remains the same, but the distance to the walls is decreased. So, they hit the inside walls of the syringe more often than before, causing the inside pressure to be greater than atmosperic pressure. When the plunger is released, the greater pressure inside the syringe pushes the plunger until it is back to its original position. At this point the inside pressure is again equal to the outside atmospheric pressure.

- In Part A, step 4, the plunger may not return to the initial volume when released due to friction between the barrel and the plunger.
- In Part A, students observe a loud popping sound when they pull the plunger from the closed syringe. You may want to have students observe the pop when they pull the plunger from an open syringe, compare this to the results with the closed syringe, and suggest a reason that the closed syringe produces the louder pop. Help students understand that the air pressure that is all around them makes it difficult to pull the plunger out of the barrel. While they are probably not usually aware of this powerful force, a normal atmospheric pressure of 14.7 pounds of force is pushing on every square inch of their bodies, as well as on the end of the plunger.
- Help students understand that the pressure of a gas is related to the number of collisions between moving gas particles and the objects in their way. Gas pressure is an average of the force acting on a given area of the container wall. The pressure inside the syringe is created by the collisions of the gas particles trapped inside it with the walls of the syringe. The pressure outside the syringe is created by the collision of the gas particles that make up the atmosphere.
- Challenge students to pose testable questions based on what they've done and learned in the lesson. They can design experiments and collect evidence to answer these questions, formulate claims about their findings, and, if time allows, present and defend their claims with their group or class. While we strongly recommend that students develop their own testable questions, you may need to seed the discussion with possible questions such as these: Does the freshness, size, color, or shape of the marshmallow make a difference in Part B? What would happen if a balloon full of water instead of air were placed in the syringe for Part B, steps 2 and 3? How would pressure changes affect foam products, such as shaving cream or hair mousse?

Explanation

Gas pressure is a measure of the average force per unit area exerted by the particles on the walls of the container. The pressure increases as the number of collisions with the walls of the container increases and, conversely, decreases as the number of collisions with the walls of the container decreases.

Gases are easy to compress because the particles that make up gases have lots of empty space between them. With a closed syringe that contains gas, pushing the plunger in decreases the volume and increases the gas's pressure. In this process, as you push the plunger in (increase the external pressure), you can actually feel the internal pressure pushing back on your hand. In the compressed syringe barrel, the distance the particles must travel before hitting the walls is reduced. (The size of the gas particles themselves does not change; only the amount of empty space between the particles changes.) Releasing the plunger at this point removes the extra external pressure you were applying to the gas. The gas expands to its original volume and its particles travel further before hitting the walls.

When you pull out the plunger, volume increases and pressure decreases. The balloon and marshmallow both expand when the pressure is lowered because the air inside them expands. (A marshmallow is principally sugar, gelatin, and water with air trapped inside.) As the marshmallow expands and shrinks, some of its trapped air escapes into the barrel of the syringe and the marshmallow eventually becomes small and wrinkled.

Answers to Student Questions

Part A

Step 1

The air being pushed out of the syringe

Step 2

a. *The air inside the syringe can typically be compressed to about 10 to 15 cc.*
b. *Upon release, the plunger returns close to the 50-cc mark.*
c. *The pressure of the gas inside the syringe was greater than the atmospheric pressure, resulting in the plunger being pushed back to its initial position where the internal pressure was equal to the external pressure.*

Step 3

a. *A hissing sound*
b. *This is the sound of air moving out of the syringe.*
c. *The air inside the syringe was at higher pressure and is moving to a region of lower pressure. (Gases move from higher pressure to lower pressure.)*

Step 4

a. *The plunger is difficult to pull.*
b. *The plunger returns close to the original (10-cc) mark.*
c. *When the plunger is released, the air that surrounds the syringe (the atmosphere) pushes the plunger back to its starting point.*
d. *The plunger does not always return to the 10-cc mark, but it does get close to it.*

Step 5

a. *A loud pop*
b. *Yes, the outside air rushes into the syringe.*

Part B

Step 2

a. *The balloon expands when the plunger is pulled out.*
b. *The balloon returns to its original volume when the plunger is released.*
c. *Since the balloon was tied shut, no gas was added to the balloon. However, the volume of air inside the balloon changed due to a decrease in pressure inside the barrel of the syringe. Evidence that no air or gas was added to the balloon is that the balloon returns to its original size when the plunger returns to its original position.*
d. *As the plunger is pulled out, the pressure of the air inside the barrel is reduced. Evidence of this is the balloon expands.*

Step 3

a. *To increase the pressure, push the plunger in.*
b. *It will be compressed.*
c. *The observations are parallel to each other. An increase in pressure causes a volume decrease and a decrease in pressure causes a volume increase.*

Step 6

a. *The marshmallow in the syringe expands so that it's bigger than the control.*
b. *The marshmallow shrinks and may even look smaller than its original size.*
c. *After repeated pulls and releases, the marshmallow in the syringe becomes smaller and more wrinkled. The control does not change.*
d. *The air in the marshmallow escaped out of the sugar-gelatin-water matrix, causing the marshmallow to shrink and wrinkle.*
e. *The marshmallow and balloon are similar in that both contain a gas that expands when the pressure applied is reduced.*
f. *The marshmallow is different than the balloon because when the gas expands in the marshmallow, a portion of the gas diffuses out of the marshmallow into the barrel of the syringe, resulting in a shriveled marshmallow. Far less gas diffuses from the relatively gas-tight, tied balloon.*

Toys Under Pressure

Overview

The engaging, hands-on explorations in Part A allow students to investigate some interesting effects of changes in gas volume and pressure. The demonstration in Part B further explores the physics and chemistry of the pressure-volume relationship of gases.

Key Concepts

- air pressure
- air takes up space
- compressibility
- gas laws
- pressure
- pressure-volume relationship of gases
- volume

National Science Education Standards

Science as Inquiry

Abilities Necessary to Do Scientific Inquiry

- *Students make observations and ask questions about how the pressure of a gas in a closed system increases when its volume is reduced. (5–8, 9–12)*
- *Students use the data they gather to construct a reasonable explanation for the behavior of air in a whoopee cushion and a closed drinking straw. (5–8)*
- *Students use evidence from the two parts of the lesson to make predictions and develop explanations. (5–8, 9–12)*

Physical Science

Structure and Properties of Matter

- *Students discover that gases are relatively easy to compress because the particles that make up gases have lots of empty space between them. (9–12)*
- *Students observe that, when the volume of the air in each system becomes compressed due to a decrease in volume, the pressure of the system increases. (9–12)*

Science and Technology

Abilities to Do Technological Design

- *Students consider how the technological design of the whoopee cushion allows it to function. (5–8)*

Part A: Student Exploration

Investigate whoopee cushions and exploding straws.

Materials

- whoopee cushion toy
- balloon
- drinking straws

Procedure

 *To avoid passing germs, use a straw to inflate the whoopee cushion.*

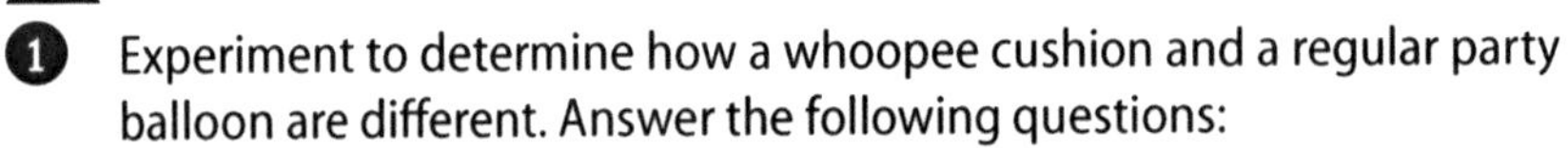

1. Experiment to determine how a whoopee cushion and a regular party balloon are different. Answer the following questions:

   a. *What holds the air inside the whoopee cushion?*

   b. *Are you able to compress the air inside the whoopee cushion even a little bit? Why or why not?*

   c. *What did you do to get the air out of the whoopee cushion?*

   d. *Describe how the whoopee cushion is engineered to provide its characteristic properties.*

2. Work with a partner to make a straw explode as follows. One partner should grip a straw tightly at both ends as shown in Figure 1a. *What's in the straw?* The other partner should sharply flick (with a finger from behind the thumb) the middle of the straw. Observe what happens.

3. With a new straw, one partner should grip the straw tightly at both ends as shown in Figure 1a. With hands positioned like pedals on a bicycle, move your hands in a pedaling motion, rolling the straw up from both ends, until about 5 cm (2 inches) of unrolled straw are left in the middle. (Do not roll the straw onto your finger. See Figure 1b.) Without letting go, observe the appearance of the unrolled portion in the middle of the straw. *Describe what you see.*

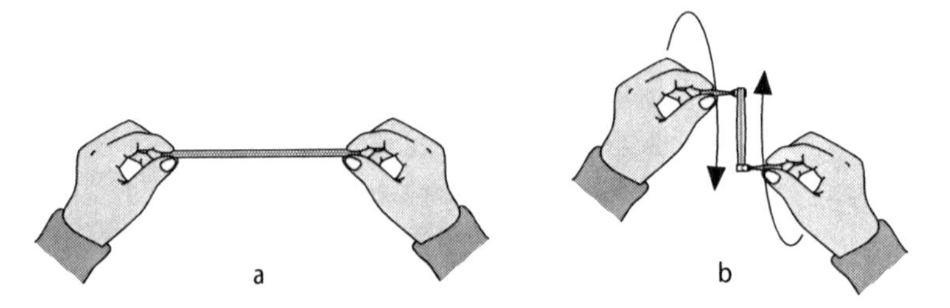

Figure 1: (a) Grasp the straw with both hands. (b) Twist one hand over another until about 5 cm (2 inches) of unrolled straw is left in the middle.

4. The other partner should sharply flick the middle of the straw. *What happens?*

5. Exchange roles with your partner, and repeat steps 2–4 with another straw.

6. Answer the following questions:

   a. *What happens to the air inside the straw when you roll the straw?*

   b. *What did you do to get the air out of the rolled straw?*

   c. *Why did the results differ when flicking an unrolled and rolled straw?*

   d. *How are the whoopee cushion and exploding straw systems similar? How are they different?*

Part B: Teacher Demonstration

How far can you blast a potato?

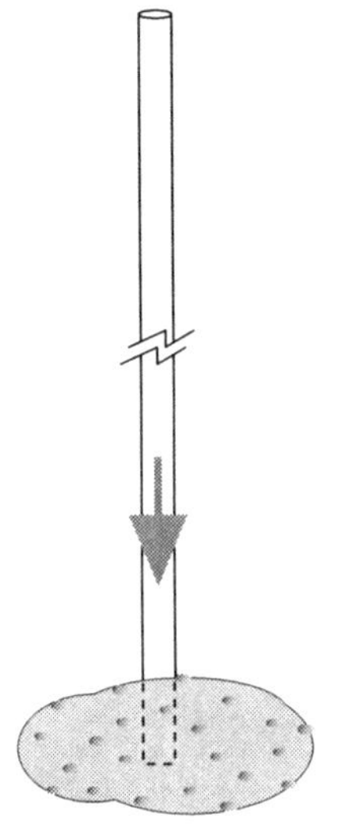

Figure 2: Push one end of the tube into the potato to plug the tube.

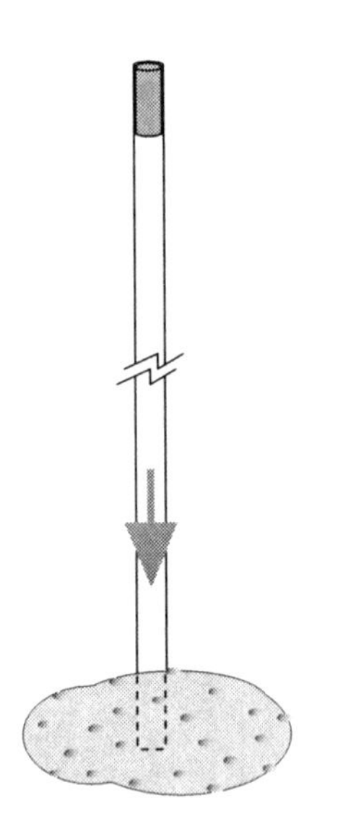

Figure 3: Plug the other end of the tube with another potato plug.

Materials

- potato
- rigid PVC, aluminum, or copper tubing, 1–2 cm (½–¾ inch) in diameter and at least 25 cm (10 inches) long
- wooden dowel that fits loosely inside the tubing (to use as plunger), about 50 cm (20 inches) long

Procedure

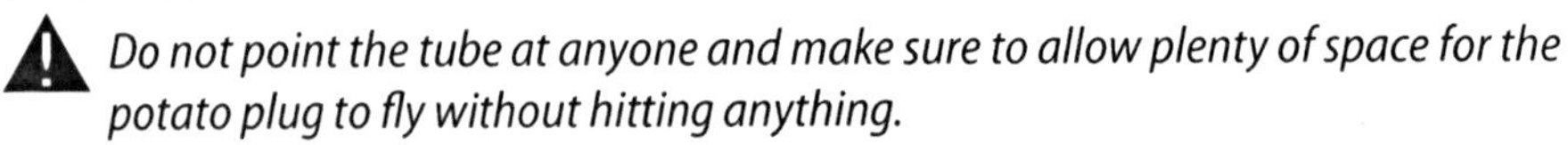

⚠ *Do not point the tube at anyone and make sure to allow plenty of space for the potato plug to fly without hitting anything.*

1. Push one end of the rigid tubing into the potato so that the end of the tube is plugged with a piece of potato. Pull the tube out of the potato. (See Figure 2.)

2. Push the other end of the tube into the potato to plug this end as well. (See Figure 3.) Pull the tube out of the potato. Use the dowel to push one of the plugs 5–6 cm (about 2 inches) into the tube. (See Figure 4.)

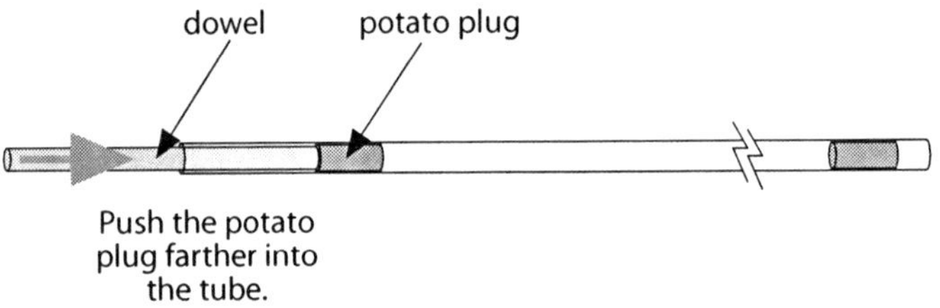

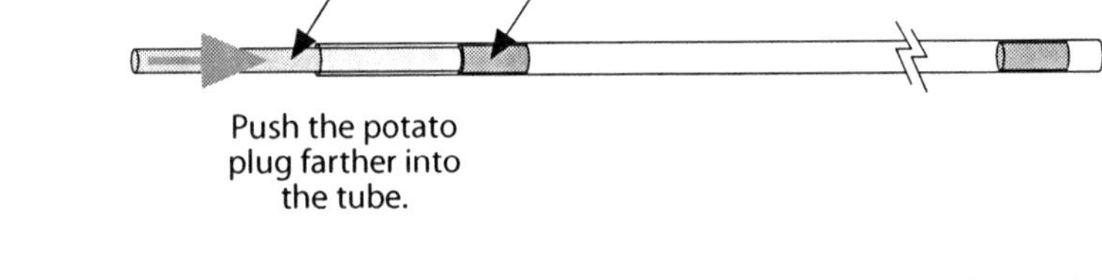

Figure 4: Push the plug 5–6 cm (about 2 inches) into the tube with the dowel.

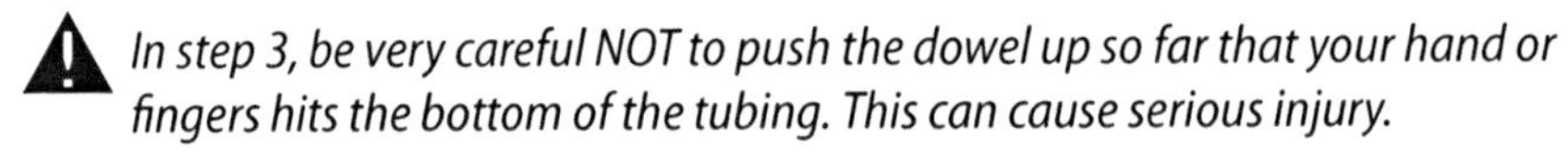

⚠ *In step 3, be very careful NOT to push the dowel up so far that your hand or fingers hits the bottom of the tubing. This can cause serious injury.*

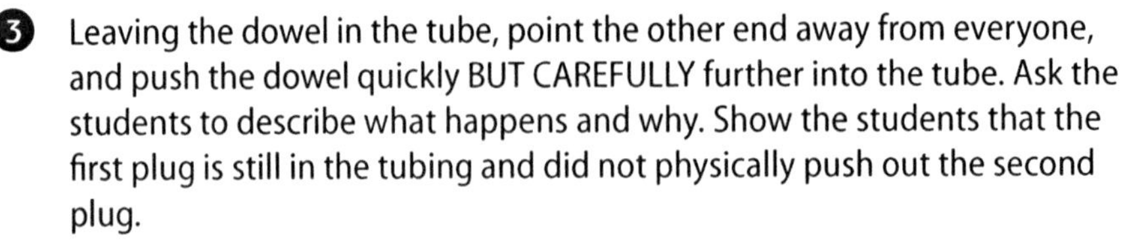

3. Leaving the dowel in the tube, point the other end away from everyone, and push the dowel quickly BUT CAREFULLY further into the tube. Ask the students to describe what happens and why. Show the students that the first plug is still in the tubing and did not physically push out the second plug.

Instructor Notes

Tips and Instructional Strategies

- Have students try to tie together what they observed in the two parts of this lesson. Stress the pressure-volume relationship and how each system has a different degree of mechanical integrity, which is breached if sufficient external force is applied to it.
- For Part A, have students suggest ideas for preventing the rolled straw from bursting, such as reinforcing the straw with tape. Give students a chance to try their ideas on additional straws.
- In Part B, be sure students understand that air is trapped between the two potato plugs.

Explanation

Part A introduces the pressure/volume relationship of a gas, known as Boyle's law. Boyle's law states that volume varies inversely with pressure when the temperature and the amount of gas are constant. Therefore, as volume decreases, pressure increases, and vice versa. Since pressure is a measure of the number of collisions with the walls of the container, reducing the volume (while temperature and number of molecules are constant) means that there are more collisions with the walls and thus there is greater pressure.

Because the opening of the whoopee cushion is a valve, if you press gently on an inflated whoopee cushion, you can slightly compress the air inside. If you press more forcefully, you breach the integrity of the system at the valve, and air rushes out, producing the toy's signature rude sound.

When the straw is gripped tightly at each end, air is trapped within the straw. As the straw is rolled up, the volume of the air decreases and its pressure increases. As a result, the unrolled middle portion bulges slightly. When "flicked," the straw breaks and the outward-rushing air makes the "pop," similar to what occurs when a balloon breaks. (Gases always move from higher pressure to lower pressure.)

In Part B, air is trapped between the two potato plugs. One plug is held in place by the dowel and cannot be moved backward. Pushing the dowel quickly inward decreases the volume of trapped air and increases its pressure above atmospheric pressure, which propels the opposite plug out of the tube with great force.

Answers to Student Questions

Part A

Step 1

a. A valve holds the air inside.

b. The air can be compressed a bit due to the large space between molecules.

c. *Sitting on the whoopee cushion or hitting it forcefully will cause the air to escape.*

d. *The valve allows air to enter the whoopee cushion. When a fairly strong force is applied, the volume of the gas decreases, which increases its pressure so that the integrity of the system is breached at the valve and the air escapes.*

Step 2

Air

Step 3

There is a slight bulge in the unrolled portion of the straw.

Step 4

The straw bursts with a sharp pop.

Step 6

a. *Air inside the straw is compressed as the volume in the middle of the straw decreases.*

b. *Flicking the straw quickly makes a hole in the straw so that air rushes out with a "pop."*

c. *Because the pressure inside an unrolled straw is no higher than atmospheric pressure, the unrolled straw does not "pop" when flicked. As the straw is rolled up, the volume of the trapped air decreases and the pressure increases, stretching the straw slightly so that the rolled straw "pops" when flicked.*

d. *The whoopee cushion and exploding straw are similar in that air is held in each system until enough force is applied to breach the system. As the volume of the trapped gas decreases, its pressure increases. (The inverse is also true.) The systems are different in that the whoopee cushion is designed to allow this breach and the straw must actually be broken to breach the system.*

Sticky Balloons

Overview

Students apply methods of inquiry to explore the effects of air pressure. The lesson can be used to pique students' interest in the behavior of gases.

Key Concepts

- atmospheric pressure
- experimental design
- gas laws
- pressure-volume relationship of gases

National Science Education Standards

Science as Inquiry

Abilities Necessary to Do Scientific Inquiry

- *Students use logic and evidence to formulate scientific explanations about why a flexible plastic cup sticks to a balloon. (5–8, 9–12)*
- *Students design and execute an investigation to hold the rigid cup to the balloon based on the evidence gathered using a flexible cup. (5–8, 9–12)*

Physical Science

Properties and Changes of Properties in Matter

- *Students observe volume changes in gases as the pressure changes. (5–8)*

Structure and Properties of Matter

- *Students refer to the relationship between the pressure and volume of gases when explaining why the cup sticks to the balloon. (9–12)*

Student Exploration

Can you make a flexible plastic cup stick to a balloon? How about a rigid cup?

Materials

- round balloon, 9 inches in diameter or larger
- flexible, transparent plastic cup, 9 ounces or larger
- rigid, transparent plastic cup, 6 ounces or larger

Procedure

1. Inflate the balloon. Squeeze the flexible plastic cup until the sides are somewhat depressed. Position the opening of the squeezed cup firmly against the balloon and slowly release the squeeze on the cup so that the cup returns to its normal shape. If the cup does not stick to the balloon, repeat the process until it does stick. With the cup stuck onto the balloon, look carefully at the edge of the balloon. Look inside the cup. *Does any of the balloon protrude into the cup? Draw a picture of what you observe.*

   How does the air pressure inside the flexible cup compare with the air pressure inside the balloon and with the atmospheric pressure outside the cup when the cup is stuck to the balloon? What is your evidence? Explain why the flexible cup sticks to the balloon.

2. Now find a way to stick the *rigid* cup to a balloon. Assume that the rigid cup will break if squeezed; that is, do not use the technique that worked in step 1. Think about why the balloon stuck in step 1 and what is flexible in this case. (You do not have to start with the balloon fully inflated.)

   Describe how you got the rigid cup to stick to the balloon. Compare how far the balloon protrudes into the rigid cup before and after you get it to stick. How do the volume and pressure of the air in the cup compare before and after you got the cup to stick? How does the pressure of the air in the cup compare with atmospheric pressure before and after you got it to stick? Explain why the rigid cup sticks to the balloon.

Instructor Notes

Tips and Instructional Strategies

- Challenge students to pose testable questions based on what they've done and learned in the lesson. They can design experiments and collect evidence to answer these questions, formulate claims about their findings, and, if time allows, present and defend their claims with their group or class. While we strongly recommend that students develop their own testable questions, you may need to seed the discussion with possible questions such as these: What effect would adding water, oil, or petroleum jelly to the rim of the cup have? Does the width of the cup's mouth or the mass of the cup make a difference? Would results differ if the activity were done at warmer or colder temperatures?

Explanation

This activity illustrates the relationship between pressure and volume of gases. Initially the pressure in the cup is equal to atmospheric pressure. In step 1, the flexible cup is squeezed and pressed against the balloon so that air cannot enter or escape. When the cup is allowed to expand back to nearly its original size (a volume increase without a loss or gain of molecules) the pressure in the cup decreases. The fact that the cup is now stuck onto the balloon is evidence that the atmospheric pressure pushing on the outside of the cup is greater than the pressure inside the cup and the fact that the balloon now protrudes farther into the cup is evidence that the pressure inside the balloon is greater than the pressure inside the cup.

When the rigid cup is placed on the partially inflated balloon, the balloon protrudes into the cup. As the balloon is inflated, the curvature lessens and there is less balloon inside the cup. Thus, the *volume* of the air inside the cup increases as the balloon is inflated, even though the *amount* of air remains constant. This causes the air pressure inside the cup to be less and, as before, the greater atmospheric pressure presses the cup to the balloon.

Answers to Student Questions

Step 1

The balloon protrudes further into the cup when the two are stuck together. When you stop squeezing the cup while it's held against the balloon, the cup returns (at least partially) to its normal shape. Since no additional air enters the cup, the increase in volume as the cup expands decreases the pressure inside the cup. The pressure inside the balloon is now greater than the pressure inside the cup; this greater pressure inside the balloon pushes the balloon farther into the cup. Even with this adjustment of pressures, the atmospheric pressure pushing on the outside of the cup and balloon is greater than the pressure inside the cup plus gravity so the cup sticks to the balloon.

Step 2

One way to stick the rigid cup to the balloon is as follows: Blow up the balloon about a third of the way. Press the open end of the rigid cup firmly into the side of the balloon. (Note the distance the balloon extends into the cup.) With the cup still in position, fully inflate the balloon and release the cup. It should stick to the balloon. When the cup is placed on the partially inflated balloon, there is more curvature of the balloon inside the cup. As the balloon is inflated, the curvature lessens and there is less balloon inside the cup. The volume of the air (gas) inside the cup is thus greater when the balloon is inflated even though the amount of air (gas) remains constant. This decreases the pressure inside the cup and thus the greater atmospheric pressure pushing on the cup makes it stick to the balloon.

Reference

Liem, T. *Invitations to Science Inquiry,* 2nd ed.; Ginn: Lexington, MA, 1987.

Degassing Soda Pop

Overview

After an engaging demonstration, students use a syringe to degas a carbonated soft drink and learn about the relationship between pressure and solubility of a gas.

Key Concepts

- air takes up space
- carbon dioxide
- experimental design
- pressure
- solubility of gases

National Science Education Standards

Science as Inquiry

Abilities Necessary to Do Scientific Inquiry

- *Students use the graduation marks on the syringe to measure quantities of air and liquid. (5–8, 9–12)*
- *Students design and conduct a scientific investigation, identifying and controlling variables and making accurate measurements. (5–8, 9–12)*
- *Students use logic and evidence to make claims about what happens to the pH of the solution as carbon dioxide is removed from carbonated water. (5–8, 9–12)*

Physical Science

Properties and Changes of Properties in Matter

- *Carbon dioxide is soluble in aqueous solutions, which gives carbonated soft drinks their fizz. This solubility is affected by pressure and temperature. (5–8)*
- *Students observe the characteristic property of pH and explore how the pH changes as carbon dioxide is removed from carbonated water. (5–8)*

Chemical Reactions

- *Students observe how pH increases as carbon dioxide is removed from carbonated water. (9–12)*

Part A: Teacher Demonstration

Shake a carbonated soft drink in a baby bottle to show the release of dissolved gas.

Materials

- plastic baby bottle
- blind nipple (baby bottle nipple without a hole) and bottle ring
- carbonated soft drink to fill the bottle

Procedure

1. Fill the bottle to the rim with the soft drink. (Pour the liquid down the side of the bottle to reduce foaming.) Securely attach the blind nipple and bottle ring to the bottle. Have students observe as you shake the bottle vigorously.
2. Ask and discuss these questions: *What do you observe? What do you think happened? Do you think it was a physical change or a chemical change? How much gas is measured in the graduated baby bottle?* Turn the bottle upside down to make it easier for students to estimate the volume of gas. (Wait for the foam to go down first.)

Part B: Student Exploration

Discover how much gas is in a fizzy soft drink.

Materials

- safe plastic syringe with 60-cc capacity and airtight rubber cap
- carbonated soft drink
- other materials as needed for the student-designed experiment in step 3

Procedure

1. Pull 20 cc carbonated soft drink into the syringe and make sure to get all the gas out as shown in Figure 1.

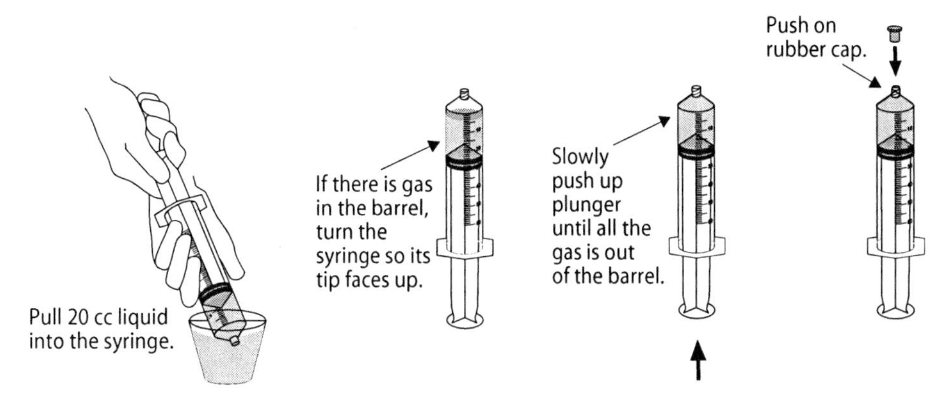

Figure 1

2. Place the rubber cap on the syringe and use your finger to hold it firmly in place. Pull the plunger back to the 60-cc mark and hold it. Shake the syringe for about 10 seconds and then release the plunger. *What happens?* Repeat this several more times. *What do you see? How much gas came out of the soft drink?*

3. Use what you've learned to design an experiment to determine the effect temperature has on the solubility of carbon dioxide in a carbonated soft drink.

Part C: Student Exploration

Further investigate the chemistry of degassing soda pop.

Materials

- safe plastic syringe with 60-cc capacity and airtight rubber cap
- club soda or unflavored carbonated water
- 2, 50- to 100-mL transparent cups or beakers
- indicator solution

➤ *If available, use methyl red indicator, as this provides the best result (a red-orange to orange-yellow color change). You can also use universal indicator, which changes from reddish pink to orange.*

- test solutions such as soapy water, baking soda and water, and vinegar
- paper towel

Procedure

1. Test the indicator to see how its color changes in an acid and a base. Try soapy water, baking soda and water, and vinegar as test solutions. Dip a piece of paper towel in the indicator solution, then drop test solutions on the paper towel to see the color change.
2. Stir about 10 drops of indicator into about 25 mL club soda or carbonated water so that the color is clearly visible. Record the color of the resulting solution.
3. Pull about 20 cc of the solution from step 1 into the syringe. Repeat Part B, step 2, making careful observations of the amount of gas released and the color of the solution after each step in the degassing process. *Based on this evidence, make a claim about how the acidity of the soda changes with degassing.*

Instructor Notes

Teacher Tips and Instructional Strategies

- Part A can also be done as a student exploration. Blind nipples (nipples without holes) can be purchased at pet stores or from online pet suppliers.
- Because different brands of carbonated soft drinks have different amount of carbonation, you may want to have students try different brands in Part B and compare the results.
- In Part B, be sure students measure the volume of gas released from the soft drink at atmospheric pressure. Students should recall that gases are compressible and expandable. The volume of gas collected will be smaller if the pressure is increased (the plunger of the syringe is pushed in), and the volume of the gas collected will be greater if the pressure is reduced (plunger is pulled out). Allow the plunger of the syringe to move freely to its equilibrium position, neither pushed in, nor pulled out. The gas will then be close to atmospheric pressure.
- In Part B, step 3, discuss experimental design as a class. Emphasize the importance of controlling variables, such as testing only one brand of carbonated soft drink while varying the temperature. Discuss how the variable of temperature could be explored, such as refrigerating or warming the soft-drink-filled syringe to targeted temperatures. You may want to develop an experimental procedure as a class and then assign groups different conditions (temperatures) to test. Alternatively, each group could develop their own procedure. If the range of temperatures tested is extreme, students may be able to detect that the amount of gas is different than in the room-temperature experiment they did originally. This observation illustrates that the solubility of a gas is lowered as the temperature of the solution increases.
- If your students' experimental design involves placing the syringe with soft drink in a boiling water bath, the rolling boil can cause water and water vapor to leak into the syringe through the syringe cap. Leakage can be minimized by placing the syringe in a plastic bag that is gathered tightly around the syringe.
- In Part C, regular soft drink will not work because it usually has citric and other acids for flavor that will mask pH changes due to the carbonic acid alone.

Explanation

In Part A, shaking the baby bottle allows the dissolved gas to be released from the carbonated soft drink. As the bottle is shaken, the amount of the gas released increases, and the nipple expands. While volume measurements may vary due to the amount of shaking and the initial temperature of the soft drink, about 200 mL gas is typically released.

The "fizz" in a carbonated beverage is caused by carbon dioxide (CO_2) gas that is dissolved in the drink. The solubility of a gas in a liquid depends on the gas pressure above the liquid and the temperature. When you open a soft-drink container, the gas pressure above the liquid is lowered and the gas starts bubbling out of solution. Putting the soft drink in the syringe and pulling the plunger lowers the pressure even more, which allows more $CO_2(g)$ to escape from the liquid.

The student-designed experiment in Part B, step 3, shows students that if they hold the volume of the carbonated solution constant while changing the temperature of the solution, the rate at which the gas is expelled and the amount expelled from the solution are related to the temperature of the solution. Solubility of a gas in a liquid decreases as the temperature of the liquid increases.

Part C illustrates the pH changes that occur when CO_2 is lost from the solution during the degassing of a carbonated water. The equilibrium is shown as follows:

$$\underset{\text{hydrogen ion}}{H^+(aq)} + \underset{\text{carbonate ion}}{HCO_3^-(aq)} \rightleftharpoons \underset{\text{carbonic acid}}{H_2CO_3(aq)} \rightleftharpoons \underset{\text{water}}{H_2O(l)} + \underset{\text{carbon dioxide}}{CO_2(g)}$$

As $CO_2(g)$ is lost from the carbonated water, the equilibrium shifts to the right, producing more $CO_2(g)$. This lowers the hydrogen ion concentration, so the pH increases. The amount of H_2CO_3 actually present in this equilibrium is small. Shakhashiri states that the ratio of CO_2 to H_2CO_3 is about 600 to 1 at 25°C.

By releasing the CO_2 from the syringe and again degassing the solution, the amount of CO_2 dissolved in the solution is further reduced. This reduces the amount of H_2CO_3, which in turn reduces the amount of bicarbonate and hydrogen ions in the solution with each decarbonation. A smaller amount of CO_2 is released from the carbonated water with each degassing cycle.

Answers to Student Questions

Part B

Step 2

a. *Gas bubbles form as the plunger is pulled outward. The gas that leaves the liquid remains in the space above the liquid.*

b. *Gas bubble formation continues, but over time the amount decreases and the volume of the gas above the liquid increases more slowly.*

c. *Students should record the volume of the carbon dioxide in the syringe.*

Part C

Step 2

The acidity of the solution decreases as the gas is removed from solution.

Reference

Shakhashiri, B.Z. *Chemical Demonstrations,* Vol. 2; University of Wisconsin: Madison, WI, 1985; pp 106–120.

Creaking Plastic Bottles

Overview

What happens to a helium-filled Mylar® balloon that sits in a car for several hours on a very cool day or on a very hot day? In this lesson, students explore similar phenomena and see how these occurrences are related.

Key Concepts

- Charles's law
- experimental design
- gas laws
- properties of gases

National Science Education Standards

Science as Inquiry

Abilities Necessary to Do Scientific Inquiry

- *Students use logic and evidence to formulate conclusions about changes in the pressure and volume of the gas in capped soft-drink bottles and plastic syringes when heated and cooled. (5–8, 9–12)*
- *Students gain experience in identifying and controlling variables as they conduct experiments to determine the volume of cooled air in the bottle. (5–8, 9–12)*
- *Students pose testable questions and design experiments to answer these questions. (5–8, 9–12)*
- *As an extension, students use Charles's law to calculate the volume of air in the bottle at room temperature and at a cooler temperature. (9–12)*

Physical Science

Structure and Properties of Matter

- *By observing the physical changes that occur in capped soft-drink bottles and plastic syringes when heated and cooled, students conclude that cooling trapped air causes its pressure and volume to decrease and heating trapped air causes its pressure and volume to increase. These conclusions are reinforced when students observe the behavior of a soap film over the mouth of an uncapped bottle when the bottle is heated and cooled. (5–8, 9–12)*

Conservation of Energy and the Increase in Disorder:

- *Students observe evidence that the higher the temperature, the greater the atomic or molecular motion. (9–12)*

History and Nature of Science

Historical Perspectives

- *Students learn about and use the mathematics of Charles's law during their explorations. (9–12)*

Part A: Student Exploration

What factors are responsible for the cracking and creaking sounds sometimes made by an empty, capped plastic soft-drink bottle?

Materials

- 2-L plastic soft-drink bottle with cap
- ice-water bath
- small plastic bottle, such as 710-mL (24-ounce) or smaller soft-drink bottle
- soapy water
- hot tap water

Avoid water hotter than hot tap water because it can cause severe burns and will permanently deform the plastic bottle, interfering with the intended outcomes of the activity. Never use a closed glass container to conduct this or any other activity that might involve a change in volume or pressure: glass explosions or implosions are extremely dangerous!

Procedure

1. Fill the empty 2-L soft-drink bottle about ¼ full with hot tap water in order to warm the air inside the bottle. Cap the bottle and swirl the water around for about 30 seconds. Discard the hot water and immediately recap the bottle. *Record any changes involving the bottle as it cools to room temperature.* Place the bottle in ice water. *Are any further changes observed? What can be concluded about the changes that accompany this cooling process?*

2. Fill the bottle ¼ full with ice-cold water in order to cool the air inside the bottle. Pour out the cold water, gently squeeze the bottle until its sides are partially collapsed, and quickly cap the bottle. Run hot tap water over the bottle to warm. *What happens? What can be concluded about the relationship between the temperature and volume of a gas?*

3. Dip the mouth of the smaller plastic bottle in soapy water so that a soap film forms across the mouth of the bottle. Place your hands around the bottle, being careful not to squeeze the bottle and change its volume. *Observe and describe the soap film.* With the soap film intact, place the bottle into the ice-water bath. *Observe and describe the soap film. Explain your results.*

   Use what you've learned in this activity to describe and explain what would happen to a helium-filled Mylar balloon if left in a hot car and if left in a cold car.

Part B: Student Exploration

What happens to the air in a syringe?

Materials

- safe plastic syringe with airtight rubber cap
- ice-water bath
- hot plate
- 500-mL glass beaker
- thermometer

Procedure

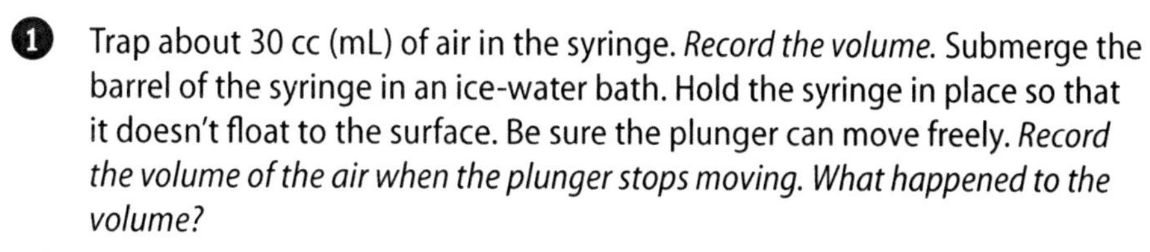

1. Trap about 30 cc (mL) of air in the syringe. *Record the volume.* Submerge the barrel of the syringe in an ice-water bath. Hold the syringe in place so that it doesn't float to the surface. Be sure the plunger can move freely. *Record the volume of the air when the plunger stops moving. What happened to the volume?*

Use caution in working with hot water, which can cause serious burns.

2. Heat a beaker of water to about 80°C (180°F). Trap about 30 cc of air in the syringe. *Record the temperature of the water and the volume of air in the syringe.* Submerge the barrel of the syringe in the hot water and hold the syringe in place, being careful to avoid touching the hot water. (Be sure the plunger can move freely.) *Record the volume of the air when the plunger stops moving. What happened to the volume?*

Instructor Notes

Tips and Instructional Strategies

- This lesson focuses on Charles's law and the relationship of the temperature of a gas to its volume and can be used to address a common student misconception. Charles's law assumes that the number of moles of gas remain constant and the pressure remains constant. In this lesson, the outside pressure on the bottle and the inside pressure of the gas remain constant, both at atmospheric pressure. In the lesson, three containers allow the volume of the gas to vary: a plastic bottle with flexible sides, a flexible soap film on a bottle, and a moveable plunger of a syringe. The student misconception occurs when, for example, a student looks at a collapsed bottle and says the internal pressure must be lower than atmospheric pressure; in fact, because the volume change occurs, the internal and external pressure remain equal.
- Due to the use of very hot water in Part B, step 2, you may want to do this step as a demonstration, especially for younger grades. As an additional safety measure, you may want to use a ring stand and clamp to hold the syringe in the hot water. Changes in temperature of a gas in a closed system can result in implosions and explosions. This activity is designed, with safety in mind, to utilize very flexible containers and only small temperature changes.
- Challenge students to pose testable questions based on what they've done and learned in the lesson. They can design experiments and collect evidence to answer these questions, formulate claims about their findings, and, if time allows, present and defend their claims with their group or class. While we strongly recommend that students develop their own testable questions, you may need to seed the discussion with possible questions such as these related to Part A, step 3: What other kinds of containers could be used that would result in observable changes in the soap film? Would a balloon over the mouth of the bottle give the same result? What is the largest container that can be used that would result in observable changes in the soap film?
- In Part B, students are deliberately instructed to "trap" the air in the syringe. This allows you to check students' conceptual understanding by verifying that students have placed the cap on the syringe. Students in older grades should understand that by trapping the air in this way, they are holding the amount of air (number of moles or "n" in the ideal gas law) constant.
- If you want to subject the air in the syringe to a 100°C (212°F) water bath for more dramatic results, the rolling boil can cause water and water vapor to leak into the syringe through the syringe cap. Leakage can be minimized by placing the syringe in a plastic bag that is gathered tightly around the syringe.
- As an extension, students in higher grades can calculate the changes in volume predicted by the gas laws and compare these to actual results. (Students may need to refer to their textbook for the mathematical expression of Charles's law.) Have students discuss any variation between experimental

and calculated volumes. Variation could be due to experimental design (for example, hot water weakening the airtight nature of the syringe or the experimenter's inability to effectively submerge the majority of the syringe barrel or to accurately read the volume while the syringe is submerged).

- Students may be interested in these real-life examples of the properties of gases: In a hot air balloon, the air inside the balloon envelope is heated, forcing some of the air out of the bottom. Since there are now fewer gas particles in the balloon envelope, there is less mass of particles, but the volume is the same. Therefore, the balloon system is less dense than the air surrounding it and the balloon will "float" on the cooler, more dense air in the atmosphere. Tire pressure is another example of the gas laws in action. Drivers are advised to measure the "cold" pressure of automobile tires for the following reason: Driving a car causes friction on the tire, increasing the temperature and thus the tire pressure. (The tire volume cannot change significantly because the rubber is reinforced with non-elastic materials.)

Explanation

The properties of a gas are described by four quantities: amount (number of gas particles), pressure, volume, and temperature. This activity demonstrates how these quantities are interrelated.

In an open container, gas molecules are free to move into or out of the container. Therefore, the number of moles of gas in the container can easily change. For example, if the container were warmed, the temperature of the gas within it would increase, which would in turn increase the average kinetic energy of the molecules. Knowing that $KE = \frac{1}{2}mv^2$ and that the mass of the molecules doesn't change, we can infer that the average velocity of the molecules would increase, causing them to strike the inside walls more often. Because the container is open, molecules of the gas would escape from the container, causing a net decrease in the amount of gas inside the container. This would result in a reduction of the number of collisions with the container walls. The pressure in this open container would adjust to be equal to that outside of the container.

In a closed container, the amount of gas (number of moles) is fixed. The nature of the container—either flexible or rigid—becomes an important factor. Just as noted above with the open container, heating a gas in a closed container increases the average kinetic energy of its molecules, which increases the rate at which the molecules in the gas phase collide with the sides of the container. If the container is closed, molecules are trapped inside. If the closed container is rigid, its volume is not able to increase. This means that molecules will hit the inside walls more often and cause the pressure of the gas inside to increase. (The inverse is also true: if there is a decrease in temperature, the pressure of the gas inside will decrease.) When the number of moles and the volume are constant, the pressure is directly proportional to the temperature in Kelvin. (This is known as Gay-Lussac's law.)

So what happens if a closed flexible container is heated? In this case, the amount of gas (number of molecules of gas) is constant, but the volume can change. As before, increasing the temperature increases the average kinetic energy of the molecules, which increases the number of collisions against the sides of this flexible container. This increased number of collisions causes the flexible sides to be pushed outward, resulting in a volume increase. The volume increases until the net pressure inside the container equals that outside the container. This phenomena is known as Charles's law, which states that the volume of a gas is directly proportional to the temperature in Kelvin if the number of moles of gas and pressure on the gas is held constant.

Answers to Student Questions

Part A

Step 1

a. *As the trapped air cools to room temperature, the sides of the plastic bottle are pushed in, decreasing the volume; this is typically accompanied by creaking sounds.*
b. *As the trapped air cools below room temperature, the bottle collapses further.*
c. *Cooling the trapped air causes the volume of the trapped gas to decrease.*

Step 2

a. *The bottle "pops" back into shape.*
b. *The volume increases as the temperature of the gas increases. (The absolute temperature and volume of a gas are directly proportional.)*

Step 3

a. *The soap film expands outward.*
b. *The soap film is pushed into the bottle.*
c. *When the air in the bottle is heated by your hands, the volume of the air increases. When the air is cooled in the ice-water bath, the volume of the air decreases. Thus, the temperature and volume of a gas are directly proportional.*
d. *The helium-filled Mylar balloon would expand and contract with heat and cold, respectively, just as the flexible containers did in this activity.*

Part B

Step 1

The volume of air in the syringe decreases.

Step 2

The volume of air in the syringe increases.

Investigations with Hand Boilers

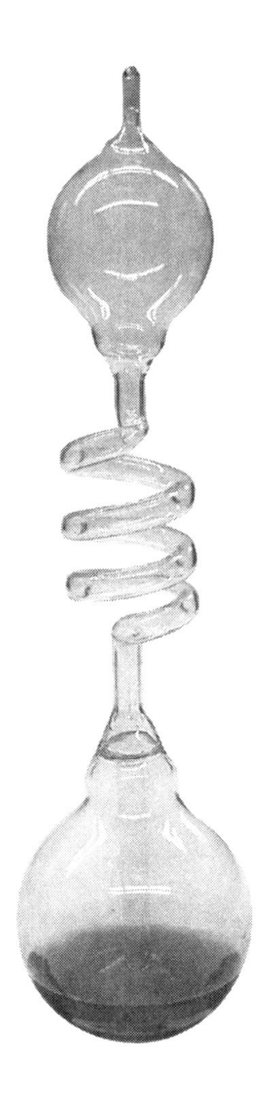

Overview

The activity begins with an exploration that challenges students to figure out how the hand boiler works. For upper grades, the lesson continues with a whole-class activity that shows students how the hand boiler can be used as a closed-system, mini-distillation apparatus.

Key Concepts

- boiling
- condensation
- discrepant events
- distillation
- endothermic processes
- evaporation
- gas laws
- phase changes
- solutes
- solutions
- solvents
- vapor
- vapor pressure
- volatility

National Science Education Standards

Science as Inquiry Standards

Abilities Necessary to Do Scientific Inquiry

- *Students describe the behavior of the hand boiler and ask questions about what causes a liquid to appear to boil when warmed by the hands. (5–8, 9–12)*
- *Students predict what is happening in the unseen lower bulb of the toy, draw their predictions, then compare these to what they observe when the lower bulb is visible. (9–12)*

Physical Science

Properties and Changes of Properties in Matter

- *Students learn that the colored liquid in the hand boiler can be separated into different components (a solvent and a solute). (5–8)*

Structure and Properties of Matter

- *Students discover that as evaporation takes place in the hand boiler, an endothermic process occurs, evidenced by the bulb feeling cold. (9–12)*

Conservation of Energy and the Increase in Disorder

- *This lesson demonstrates that the ethanol vapor in the hand boiler expands due to the heat from the hand. (9–12)*

Science and Technology

Abilities to Do Technological Design

- *Students learn that distillation is a technological process used for purification. (9–12)*

Part A: Student Exploration

How does this toy work?

Materials

- at least 2 hand boilers per class (ideally 1 per group)
- warm tap water
- ice water
- ice
- cups

Procedure

 *Handle the hand boiler with care; the glass is fragile.*

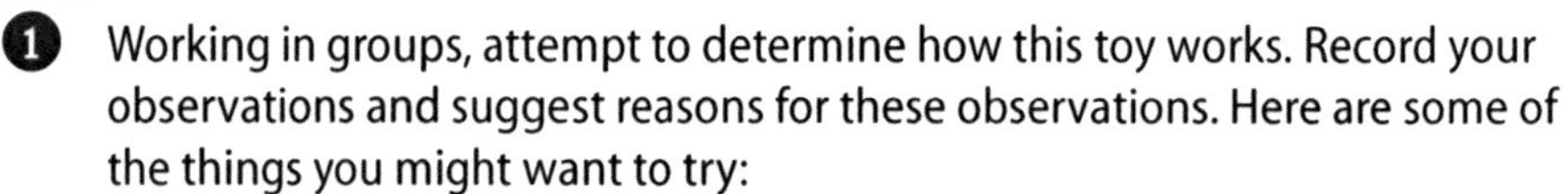

1. Working in groups, attempt to determine how this toy works. Record your observations and suggest reasons for these observations. Here are some of the things you might want to try:

   a. Compare what happens when the toy is standing on the table without being touched to what happens when you pick it up and hold it in the upright position with the palm of your hand wrapped around the lower bulb.

   b. Experiment to determine if the orientation of the toy makes a difference as to how it works. Try turning it sideways and upside down.

   c. Compare what happens if you place the bottom bulb in a cup of lukewarm tap water, room-temperature water, and ice water.

   d. Get all the liquid to the bottom bulb and stand the toy on the table. Hold a piece of ice on the top bulb without touching the bottom bulb.

2. Look closely at the construction of the toy. Draw a picture of the system and add arrows to describe the movement of the liquid you observed in step 1a. *Do you think the liquid in the hand boiler is boiling? Why or why not?*

   How do you think the hand boiler works? Suggest a new name for this toy based on what you've learned about how it works.

Part B: Whole-Class Activity

What other cool thing can we do with this toy?

Materials

- 2 hand boilers
- Styrofoam® cups
- crushed ice
- rock salt or table salt (NaCl)

Procedure

1. Carefully observe as your teacher sets up a hand boiler for this part of the lesson. Draw a picture of the setup once it is in place. Your teacher will:

   a. Transfer as much of the colored liquid as possible back to the larger bulb of the hand boiler. Then, turn the hand boiler upside down as shown at top left.

   b. While securely holding the larger bulb, place the empty smaller bulb into a tall Styrofoam cup and add alternating layers of ice and salt. This is setup 1. (See photo.)

   c. Do a second setup, but this time do not invert the hand boiler. This is setup 2. (See photo.)

Setup 1

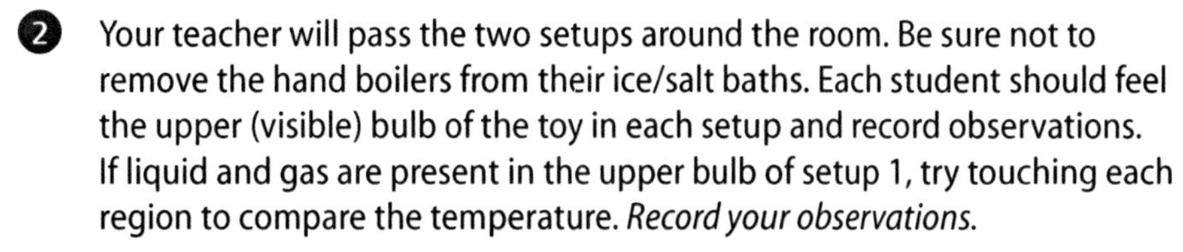

2. Your teacher will pass the two setups around the room. Be sure not to remove the hand boilers from their ice/salt baths. Each student should feel the upper (visible) bulb of the toy in each setup and record observations. If liquid and gas are present in the upper bulb of setup 1, try touching each region to compare the temperature. *Record your observations.*

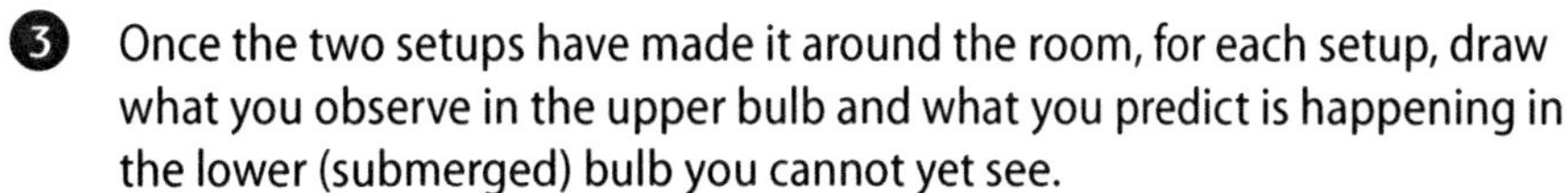

3. Once the two setups have made it around the room, for each setup, draw what you observe in the upper bulb and what you predict is happening in the lower (submerged) bulb you cannot yet see.

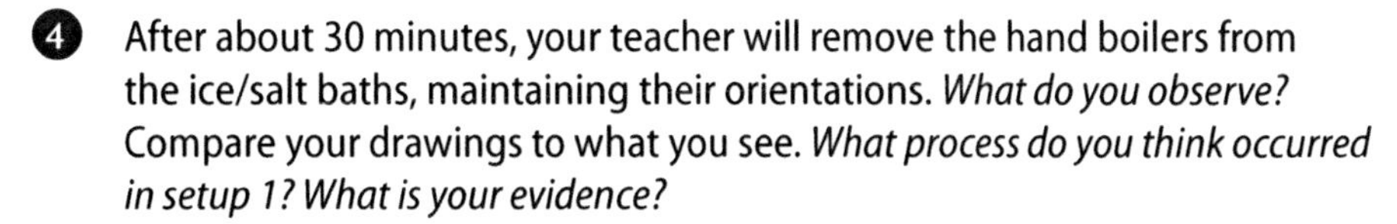

4. After about 30 minutes, your teacher will remove the hand boilers from the ice/salt baths, maintaining their orientations. *What do you observe?* Compare your drawings to what you see. *What process do you think occurred in setup 1? What is your evidence?*

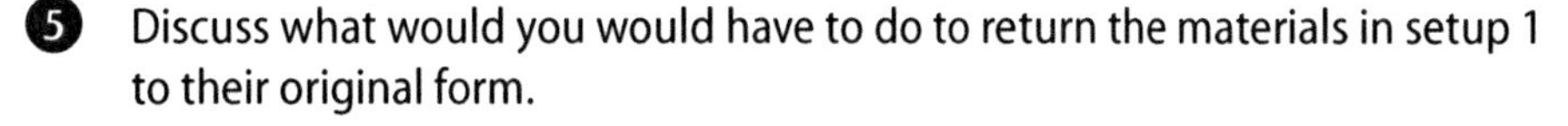

5. Discuss what would you would have to do to return the materials in setup 1 to their original form.

Setup 2

Instructor Notes

Tips and Instructional Strategies

- Students should warm the boiler using only their hands or lukewarm (*NOT HOT*) tap water. Because the boiler is a closed system made of thin glass, it could shatter if warmed to too high a temperature. Do not heat the hand boiler directly with or use it near a flame.
- Hand boiler toys come in various shapes and are sometimes sold as a Passion/Love Meter.
- To transfer as much of the colored liquid as possible into the larger bulb of the hand boiler for Part B, turn the hand boiler to the upright position and warm the smaller bulb with your hands until the liquid is forced down into the larger bulb. It's okay if a tiny amount of liquid remains in the smaller bulb.
- Older students continue to explore using the hand boiler as a distillation apparatus in Part B. When students observe that the top (larger) bulb is cold in setup 1, they should know that it is the contents of the upper bulb (vapor and liquid) that are cold and not the entire apparatus getting cold through conduction from the ice/salt bath. (The glass itself is not a very good conductor of heat.) The coldness of the upper bulb should be a clue that the evaporation of the liquid caused the cooling.
- At the end of Part B, you may need to lead students to understand the process that occurred and to identify this process as distillation. You may also want to share with them that the distillation process has many industrial applications. For example, gasoline and motor oil, which come from crude oil, are products of distillation.

Explanation

The colored liquid in the hand boiler is typically ethanol (the solvent) with a solid dye (the solute) dissolved in it. (Note: Some hand boilers may contain methylene chloride as the solvent.) Warming the liquid and gas contained in the toy causes the observed phenomenon. If you rest the bottom bulb in the palm of your hand, heat from your hand is transferred to the liquid, which increases its vapor pressure. This forms more gaseous ethanol, which increases the pressure inside the bottom bulb, and some of the liquid is pushed up the inner tube. As the temperature of the gas increases, its pressure increases and even more liquid is pushed up the tube. When the level of the liquid in the bottom bulb drops below the end of the tube, the gas travels up the tube and rises through the liquid, giving the appearance of boiling. If you hold the bottom bulb with your hand encircling the bulb, both the liquid and the gas above it are warmed, resulting in more dramatic bubbling in the toy. The same phenomenon is observed when the smaller top bulb is cooled with a piece of ice. This decreases the pressure in the top bulb by lowering the temperature of the gas, which means that the higher pressure in the bottom bulb will force liquid up the tube.

In Part A, students are likely to conclude that the liquid is actually boiling because they see bubbles of vapor rising through the liquid in the upper bulb. However, close observation shows that the bubbles result only after all of the liquid above the tube in the lower bulb has been pushed into the top bulb. Since the gas is still at a higher pressure in the bottom bulb, it is forced into the tube and rises up through the liquid in the upper bulb. The bubbles are not forming within the liquid itself as would be the case with boiling.

Distillation is a common means of separating components of a solution. For example, if a solution has one component that is more volatile (it boils at a lower temperature) and one that is less volatile, distillation can separate the two components. In this activity, the dyed ethanol solution contains volatile ethanol and nonvolatile solid dye. When the solution is heated, the volatile component vaporizes and passes through a tube, where it is cooled and allowed to condense back into its liquid state. The nonvolatile component remains behind in the original chamber and is not transferred in the process.

In Part B, when the empty smaller bulb is placed in the ice/salt bath, the low temperature causes some of the vapor in the smaller bulb to condense to the liquid phase. With fewer molecules in the gas phase in the bottom (smaller) bulb, the pressure in the bulb decreases. The resulting difference in pressure between the two bulbs causes more liquid to vaporize in the top (larger) bulb. Molecules of gas in the bottom bulb continue to cool and condense. Since vaporization is an endothermic process, the remaining liquid in the upper bulb is cooled. The process continues until all the liquid has evaporated and most has moved from the top bulb to the bottom (smaller) bulb. Since the dye used to color the liquid is not volatile, it remains in the top bulb.

When the process is complete, the upper bulb contains only the solid dye. The bottom bulb contains liquid ethanol. Careful observation during the distillation process shows that the intensity of color in the upper bulb is increasing since the concentration of dye is increasing as the volume of ethanol is decreasing. Likewise, since the dye is not volatile, the vapor is pure ethanol, and when it condenses, it will be colorless. In this activity, the ethanol condensed into the smaller bulb is usually slightly colored because some of the original colored solution coats the tube connecting the two bulbs.

Answers to Student Questions

Part A

Step 1

a. *With your hand wrapped around the lower bulb, the liquid moves up into the top (small) part of the toy and bubbles form. It looks like it's boiling.*
b. *The liquid is not transferred when the hand boiler is sideways or upside down.*
c. *In lukewarm water, the liquid quickly moves up into the top part of the toy and the bubbles form more rapidly. In room-temperature water and ice water, the liquid in the larger bulb doesn't bubble or move up into the top bulb.*
d. *The liquid moves up into the top bulb just like it did when the bottom bulb was held in the palm of a hand.*

Step 2

a. *It couldn't be boiling due to heating the liquid, because the liquid doesn't form bubbles (boil) when the liquid-filled bulb is held sideways or upside down. Also, touching the top bulb with ice makes the liquid go up into the top bulb just like it did when the bottom bulb was held in the palm of a hand.*
b. *The toy works because you heat the gas with your hand, the gas expands, and the liquid is pushed into the upper bulb. Bubbles occur when the gas moves up the tube into the liquid.*

Part B

Step 2

The upper bulb in each setup feels cooler than room temperature, and the upper bulb in setup 1 (where the liquid and gas are present) is much colder than the upper bulb in setup 2, which contains only gas.

Step 4

a. *In setup 1, the upper bulb has little or no liquid. A residual solid is present. The bottom bulb now contains most or all of the liquid, which is colorless or faintly colored. In setup 2, the top bulb got colder and the liquid in the small bulb is still there and still colored.*
b. *Distillation*
c. *Evaporation then condensation separated the nonvolatile dye from the volatile alcohol.*

References

David W. Brooks Teaching and Research Website. chemmovies.unl.edu/chemistry/beckerdemos/BD055 (accessed Mar 2009), Expt 055—Closed Distillation Apparatus.

Sarquis, J.L.; Sarquis, A.M. Toys in the Classroom, *J. Chem. Educ.* **2005,** *82* (10), 1450–1453.

Boiling in a Syringe

Overview

Students observe water and other liquids boiling at temperatures below their normal boiling points because the pressure has been reduced.

Key Concepts

- boiling
- boiling point
- effects of pressure on boiling point
- normal boiling point
- phase changes
- physical changes
- vapor pressure

National Science Education Standards

Science as Inquiry

Abilities Necessary to Do Scientific Inquiry

- *Students identify questions that can be answered through scientific investigations. (5–8, 9–12)*
- *Students conduct a scientific investigation, controlling variables and making accurate measurements. (5–8, 9–12)*

Physical Science

Properties and Changes of Properties in Matter

- *The boiling point of a substance is dependent on the pressure it experiences. (5–8)*

Structure and Properties of Matter

- *The intermolecular forces that bind molecules to the surface of a liquid are overcome during evaporation. Boiling occurs when liquid is converted to a vapor within the liquid as well as on its surface. (9–12)*

Conservation of Energy and the Increase in Disorder

- *As temperature increases, molecules become more energetic. (9–12)*

Student Exploration

Can you solve the mystery of the bubbles?

Materials

- thermometer
- safe plastic syringe with 60-cc capacity and airtight rubber cap
- hot, room-temperature, and ice-cold water in cups or other containers
- paper towels for cleanup

Procedure

1. Measure and record the temperature of the hot water. Fill the syringe with 20–30 cc (mL) hot tap water. (See Figure 1.) Dislodge any air bubbles that might be in the syringe by holding it with the tip up as shown in the figure and tapping it with your fingers. Push the plunger up until all the air is out of the barrel. Place the rubber cap on the tip.

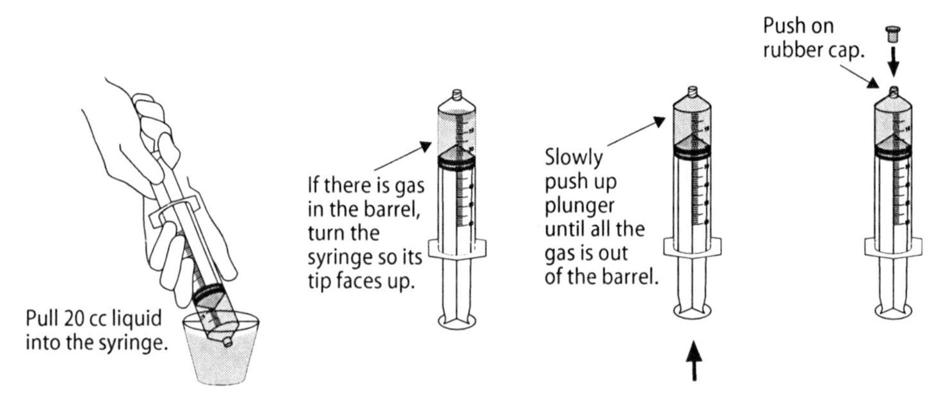

Figure 1

2. Hold the rubber cap on the syringe with your finger. Try compressing the water. *What happens?*

3. While still holding the rubber cap, pull the plunger back to the 60-cc mark and hold it there. Shake the syringe. *What happens? Does the volume of the water change?* Release the plunger, and then pull it outward and shake the syringe as before. *What happens?* When you're done, squirt the water out of the syringe.

4. Try using room-temperature water and then ice-cold water. Be sure to measure and record the temperature of the water. *Compare your observations to those when you used the hot water.*

Instructor Notes

Tips and Instructional Strategies

- This lesson is most effective if students have a basic understanding of normal boiling points and the boiling process. It can also be used to introduce or reinforce the concept that the temperature at which a liquid boils is dependent on pressure.
- Some students might suggest that the bubbles they see are actually dissolved gases, such as the nitrogen and oxygen in the air, being released. This would certainly account for some of the behavior, but not all of it. To avoid any doubt, you can boil the water to remove any dissolved gases before it is used.
- Through discussion, bring out the idea that pressure and boiling point are related. Remind students that the temperature of the hot water they used was far lower than the normal boiling point of water. Ask students what conclusion they could draw about the relationship between pressure and boiling point based on the results of the exploration. Make sure that students grasp the idea that a liquid's boiling point decreases as the pressure above the liquid decreases.
- Connect what students learn about boiling to the concepts of vapor pressure and evaporation. You may want to conduct an "evaporation race" by streaking a blackboard with water and with rubbing alcohol. Based on what they've learned, you might ask students to identify the liquids based on their rate of evaporation or to predict the relative rates of evaporation. If a blackboard is not available or to further reinforce the concept through a tactile experience, use cotton balls to dot the students' arms with alcohol and water so that they can predict the identity of each liquid based on the relative rates of evaporation.
- Because of the volatility of alcohol, you may want to demonstrate boiling in the syringe with rubbing alcohol. Students will observe more vigorous boiling with a smaller pressure change. However, note that the markings on the syringe are apt to dissolve off if exposed to alcohol.

Explanation

Boiling and evaporation are both examples of a phase change. Evaporation occurs when molecules on the surface have sufficient energy to overcome the attractive forces exerted on them by neighboring molecules. Thus they can escape from the bulk liquid into the gas phase. Evaporation occurs at any temperature; however, the rate of evaporation depends upon the temperature. Molecules in the liquid phase have different amounts of energy. The higher the temperature, the higher the average energy. Since only the highest energy molecules can escape, a higher temperature means a larger fraction of the molecules can evaporate. Thus evaporation occurs more rapidly as the temperature increases. However, liquid at any temperature left in an open container will eventually all evaporate. (For a substance with a very, very low vapor pressure, this can be an exceptionally slow process.)

Boiling occurs when the vapor pressure of the liquid is equal to the confining pressure. If a liquid is placed in a closed container, some of the liquid evaporates. As more and more liquid is converted to the gas phase, the likelihood of a molecule in the gas phase colliding with the liquid surface and re-entering the liquid increases. This is condensation—the reverse of evaporation. When the rates of evaporation and condensation are the same, no more liquid appears to evaporate. However, if individual molecules could be observed, each molecule converted into the gas phase would be balanced by a molecule in the gas phase being converted to the liquid phase. This type of phenomenon is called a dynamic equilibrium. The pressure of the gas particles when equilibrium is reached is the vapor pressure. The vapor pressure of a liquid depends upon the temperature of the liquid; the higher the temperature, the higher the vapor pressure.

The confining pressure of a liquid (in an open container exposed to the atmosphere) is the atmospheric pressure. Since the normal pressure at sea level is 1 atmosphere (atm) or 760 torr, the normal boiling point of a liquid is defined as the temperature at which the liquid has a vapor pressure of 1 atm. For water, the normal boiling point is 100°C (212°F) and its vapor pressure is 760 torr. If the pressure is less than 1 atm, the boiling point of water will be less than 100°C. In this lesson, the pressure is decreased within the syringe when the plunger is pulled outward. The water then boils when its vapor pressure is equal to the reduced pressure in the syringe, and that occurs at a temperature lower than 100°C. This is comparable to the effect that occurs at very high altitudes. For example, the boiling point of water in Denver, Colorado (which is at 1.6 km or 5,280 feet above sea level) is about 95°C (202°F) since the atmospheric pressure in Denver is less that 1 atm. For this reason, the time to soft-boil a "3-minute egg" in Denver is actually 4 minutes since the temperature of the boiling water is lower.

If the external pressure is greater than 1 atm, the boiling point of water is greater than 100°C. For example, pressure cookers speed up the cooking process by increasing the pressure to more than 1 atm. The increase in pressure causes water to boil at a higher-than-normal temperature (above 100°C), and thus shortens the time needed to cook the food.

Answers to Student Questions

Step 2

Nothing observable happens; water, a liquid, cannot be significantly compressed.

Step 3

a. Bubbles form on the plunger and the walls of the syringe. Some of the bubbles break from these surfaces to the top of the water.

b. No, the liquid volume does not change after the plunger is released. (If the water contained dissolved gases, then that gas will remain.)

c. Bubbles form again and shaking the syringe dislodges bubbles, which rise to the surface.

Step 4

Little difference is observed with the water at different temperatures.

That Cold Sinking Feeling

Overview

Students discover that the density of liquid water varies with temperature.

Key Concepts

- density-temperature relationship
- mixing
- relative density
- temperature
- water and its properties

National Science Education Standards

Science as Inquiry

Abilities Necessary to Do Scientific Inquiry

- *Students develop descriptions and explanations for the process of mixing hot and cold water. (5–8)*
- *In the assessment, students logic and evidence to make predictions about the relative densities of hot and cold water compared to balloons filled with room-temperature water. (5–8, 9–12)*
- *As an extension, students apply logic and evidence regarding what they know about the density of water at different temperatures to explain the turnover of water in lakes. (5–8, 9–12)*

Physical Science

Properties and Changes of Properties in Matter

- *Students discover that water has characteristic properties, such as density, but that the density varies with temperature. (5–8)*

Transfer of Energy

- *Students observe the mixing that results as hot and cold water adjust to room temperature. (5–8)*

Structures and Properties of Matter

- *Students learn about the properties of water, including its density-temperature relationship, and connect these characteristics to the intermolecular forces among water molecules. (9–12)*

Teacher Demonstration

Will water always mix?

Materials

- 4 clear, colorless glass bottles of identical size

The wider the lip or the thicker the glass, the easier it will be to balance the bottles top-to-top. Wide-necked salad-dressing bottles or gas-collecting bottles are good choices. (See Figure 1.)

- hot tap water (enough to completely fill 2 bottles) colored red with food color

Do not boil the water as this may cause glass containers to break. The water should be below 60°C (140°F) to avoid scalding yourself or breaking the glass.

- ice-cold water (enough to completely fill 2 bottles) colored blue with food color

Prepare ice-cold water by placing ice cubes in a container of cold water. Remove ice cubes before filling the bottles.

- 2 index cards
- stirring rod or other utensil to mix food color into water

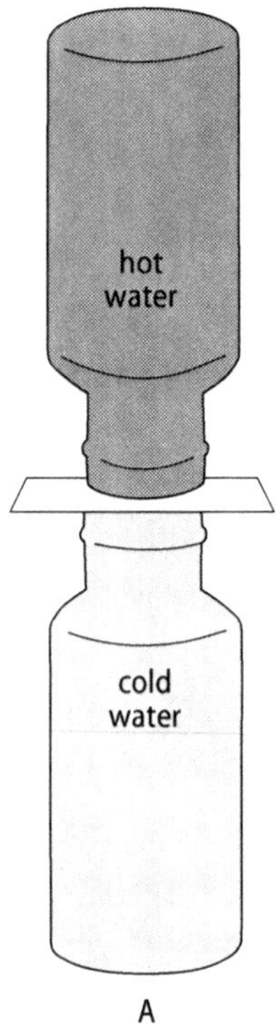

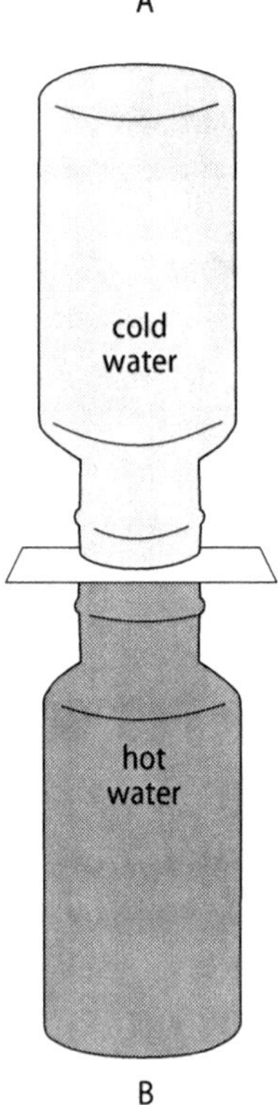

Figure 1: Set up two sets of water-mixing bottles as shown.

Procedure

1. Pour the hot water into two bottles and the cold water into the other two bottles.
2. Cover one hot-water bottle with an index card. Holding the card firmly against the mouth of the bottle, carefully invert the bottle and rest it on the mouth of a cold-water bottle, taking care not to spill the hot water on yourself. The card will be between the mouths of the two bottles. (See setup A in Figure 1.)
3. Repeat step 2, this time covering and inverting a cold-water bottle and placing it on a hot-water bottle. (See setup B in Figure 1.)
4. Have students predict what will happen in each system after the index card is removed. Have students use diagrams similar to Figure 1 to draw what they predict will happen after the cards are removed.
5. Carefully remove the index card from setup A and then from setup B.
6. Have students record their observations for the two setups. Challenge them to provide evidence that mixing may or may not be occurring. Have them predict what would have to happen for a change to occur in setup A and how long it may take. If possible, leave this setup for students to view later in the class period or the next day so that they can check their predictions.

Instructor Notes

Tips and Instructional Strategies

- Tell students that the red water is hot and the blue water is cold. Explain that the food color is added only to show the location of the hot and cold water and does not affect the results.
- You may want to set the bottles in a tray or clear plastic tub to contain spills. After removing the index cards, prop white paper behind the setups for easier viewing. For a more dramatic effect, you may want to work with a student volunteer to simultaneously remove the index cards.
- After removing the cards, change should occur immediately in the hot-water-bottom system (setup B). In the cold-water-bottom system (setup A), mixing often takes more than 30 minutes to become noticeable. Complete mixing may take several hours.
- You may want to challenge older students to draw diagrams to represent the water molecules in this activity in order to explain their observations.
- Have students look up the density of liquid water at various temperatures (or show them Table 1 in the Explanation). Challenge them to use this data to explain the observations they made in the demonstration.
- You may want to conduct an assessment of this lesson as follows:
  1. Fill two small water balloons or 3-inch regular balloons with equal amounts of water. It is necessary to fill them at a tap as the water pressure must be sufficient to stretch the balloons to a diameter of several inches. Be sure that the balloons are completely full of water with no air bubbles. Allow the balloons to equilibrate to room temperature.
  2. Prepare large, clear containers of ice-cold water and hot tap water. (Pop-beakers made from cut-off 2-L bottles work well.)
  3. Have students predict what will happen when the balloons containing room-temperature water are put into the ice-cold water and hot water and explain the predicted behavior in writing and/or diagrams.
  4. Put the balloons into the ice-cold and hot water and observe. (The balloon in the ice-cold water will float because it is less dense than the ice-cold water. The balloon in the hot water will sink because it is more dense than the hot water. The density of liquid water decreases as its temperature increases.)
  5. Have students make a claim about the system that they can substantiate with evidence they have observed.
- As an extension, you may want to discuss how this activity relates to turnover of the water in lakes or have students research this phenomenon.

Explanation

Density relates the mass of an object to its volume. Each type of matter has a characteristic density at a given temperature. In this activity, the relative densities of liquid water at two different temperatures are indicated by the movement of colored water.

In setup A, in which the hot water is placed on top of the cold water, little or no change is initially observed. The hot-on-top-of-cold system remains a two-layer system for quite some time. The more dense cold water does not rise through the hot water. Eventually, when the temperatures of both water layers reach room temperature and the density of the layers is the same, the two layers begin to mix. In setup B, the cold-on-top-of-hot system, mixing occurs almost immediately. The more dense cold water sinks through the less dense hot water, which is itself rising upward.

Water is quite unusual for a liquid. While it becomes more dense as it cools (as most liquids do), when cooled below 4°C (39°F), it expands slightly and becomes less dense. When water freezes, it expands even further. This is due, in part, to the formation of hydrogen bonds in water. Thus, ice floats in liquid water. It is very unusual that a solid is less dense than its liquid form. See Table 1 for the density of water at various temperatures.

Table 1: Density of Water at Various Temperatures (g/cm^3)*

water, ice, 0°C (32°F)	0.9167
water, liquid, 0°C (32°F)	0.99984
water, liquid, 4°C (39°F)	0.99998
water, liquid, 10°C (50°F)	0.99970
water, liquid, 20°C (68°F)	0.99821
water, liquid, 30°C (86°F)	0.99565
water, liquid, 40°C (104°F)	0.99222
water, liquid, 50°C (122°F)	0.98803

*Excerpted from *CRC Handbook of Chemistry and Physics*, 89th Edition, 2008

The expansion of water as it freezes has a significant economic impact. When water gets into cracks in roadways, freezing causes the water to expand, which can cause buckling of the road surface. Repetition of the freeze-thaw cycle causes bigger cracks and then potholes to form. This is an especially serious problem in cold-weather states during the winter.

References

Sarquis, A.M., Sarquis, J.L., Eds. Density Water Fountains. *Fun with Chemistry: A Guidebook of K–12 Activities*, Vol. 1, 2nd ed.; Institute for Chemical Education: Madison, WI, 1995; pp 133–137.

Sootin, H. *Experiments with Water*; Grosset & Dunlap: New York, 1971.

Taylor, B. *Water at Work*; Franklin Watts: New York, 1991.

Willow, D.; Curran, E. *Science Sensations*; Addison-Wesley: Reading, MA, 1989.

Visualizing Matter

Overview

This lesson helps beginning students visualize the chemical concepts of pure substances and mixtures and comprehend some of the specialized language that chemists use. Students will also visualize the reason for and process of formula writing.

Key Concepts

- atoms
- chemical bonds
- chemical formulas
- chemical reactions
- compounds
- elements
- formula writing
- isomers
- mixtures
- models
- molecular formulas
- molecules
- particle nature of matter
- pure substances
- structural formulas

National Science Education Standards

Science as Inquiry Standards

Abilities Necessary to Do Scientific Inquiry

- *Students use physical models of atoms and compounds to compare and contrast the physical and chemical properties of the materials they represent. (5–8, 9–12)*
- *Students use physical models of atoms and molecules to illustrate the differences between mixtures, pure substances, elements, and compounds, and the concept of an isomer. (5–8, 9–12)*

Physical Science

Properties and Changes of Properties in Matter

- *Using pop-beads and balls as visual aids, students learn that elements combine to form compounds and that mixtures can be made of various combinations of elements and compounds. (5–8)*
- *Students learn that mixtures can be separated by physical processes, but compounds can only be separated by undergoing a chemical reaction to break the chemical bonds. (5–8)*

Structure of Atoms

- *The toys used in the lesson illustrate that matter is made up of minute particles called atoms. (9–12)*

Structure and Properties of Matter

- *Students learn that an element is composed of a single type of atom. (9–12)*
- *Students see that a compound is formed when two or more kinds of atoms bind together chemically. (9–12)*

- *Students learn that a single molecular formula can represent more than one physical arrangement of atoms and that these resulting isomers have different chemical and physical properties. (9–12)*

Chemical Reactions

- *Students learn that breaking compounds into their constituent elements requires energy to break the chemical bonds. (9–12)*

Part A: Teacher Demonstration

Visualize elements, mixtures, and compounds.

Materials

- set of ball models in zipper-type bags prepared as described in Instructor Notes

➤ *Models can be made with balls and Velcro fasteners.*

Procedure

1. Show students the bags of balls. Review with students that matter consists of pure substances and mixtures. Review the definition of each term. Tell students that their first task is to determine whether the contents of each bag represent a pure substance or a mixture. Additionally, if a bag contains a pure substance, they need to decide if it is a monatomic element, diatomic element, or a compound. Ask students if they can complete this task by looking at the "matter" in the closed bags. (Lead them to understand the importance of opening the bag and investigating the individual pieces.)

2. Hold the bag of monatomic elements up so the students can see it, but do not reveal what the contents represent. Take the balls out of the bag one at a time. Ask the students to describe what they see. Are the balls the same or different? Do these balls represent a pure substance or mixture? Do they represent an element or a compound?

3. Repeat step 2 with the mixture set.

4. Repeat step 2 with the compound set.

5. Repeat step 2 with the diatomic element set. Does this set contain a compound? Why? What's the difference between this set and the set observed in step 2?

Part B: Teacher Demonstration

Discover the difference between separating a mixture of elements and separating a compound into elements.

Materials

- set of ball models from Part A

Procedure

1. Select two students to assist with this part of the lesson while the rest of the class is instructed to observe carefully and judge which of the volunteers completes their task first. Give one student the mixture of elements bag and the other student the compound bag.
2. Tell your two assistants to open their bags and separate the balls into two piles (by color or pattern) as quickly as possible.
3. Stop the process as soon as the first assistant is finished. Which assistant finished the task first? Why? Ask the class to explain how this student accomplished the task.
4. What did the student with the compound set have to do to accomplish the task? Ask the students what the Velcro represents in the model. Discuss the fact that energy is required to break bonds. Compare the ease of separating the parts of the mixture to that of separating the atoms of a compound.
5. Emphasize the fact that the compound has a different set of properties than the elements from which it was made. Compare this to the separation of the mixture, which *did not* chemically change the components.

Part C: Student Exploration

Apply what you've learned about elements, mixtures, and compounds while learning to write chemical formulas.

Materials

- sets of pop-bead models in zipper-type bags prepared as described in Instructor Notes

Procedure

1. Complete the worksheet (provided) using the eight sample bags labeled A–H. Each sample bag contains several pop-beads; some are joined and some are not. Leave the bags and beads as they are, because you need to observe them just as you find them.

➤ *Some sample bags contain several different colors and shapes of pop-beads, while others may contain only one type. Within each sample, any two pop-beads that are exactly alike represent the same element. When two pop-beads are joined together, this means there is a bond between the "atoms" (similar to the Velcro in Parts A and B).*

2. Compare the results of step 1 with others in the class. Discuss possible reasons for differences that might occur.

3. Look at sample bag C. (Do not remove the contents.) *Is there a way to arrange the three colored beads that would yield a different compound?* What would this arrangement look like? Sample bag P is a random collection of pop beads not meant to represent any particular sample of matter. Open it, and build a model to represent this new compound. Compare this model to the contents of sample bag C. *For this new molecule, write a chemical formula in such a way that the reader would not confuse it with the molecule represented in bag C but would know that it is composed of the same atoms.* Disassemble your new molecule and return the pop beads you used to sample bag P.

4. Look at sample bag E. (Do not remove the contents.) Write structural formulas to describe the molecules in this bag.

5. Use the pop-beads in bag P to build a set of three or more molecules that are all isomers of each other. Pass your molecules to your neighbors and challenge them to write the molecular and structural formulas for these molecules. Check their responses against the molecules you made.

Student Worksheet			
Sample Letter	Pure substance or mixture?	Elements(s) or compound(s)?	Formula for each substance represented in the bag
A			
B			
C			
D			
E			
F			
G			
H			

Instructor Notes

Tips and Instructional Strategies

- It is important to point out to the students that the balls used in this activity do not behave in the same ways as real atoms and molecules. Explain that you are using a model to help students conceptualize the actual chemical principles involved. Ask the students to discuss the limitations of using models.
- For Parts A and B, assemble sets of balls using Velcro fasteners to connect balls of different colors or patterns as shown. Place sets into zipper-type bags. Do not label the bags.

monatomic element set

diatomic element set

mixture set

compound set

- For Part C, create eight sample bags of pop-beads per set. Label the sample bags only with the letters A through H. The photos below show examples of how the sets of beads should look and the corresponding molecular and structural formula.

A: Pure Substance
Compound
(XY; X-Y)

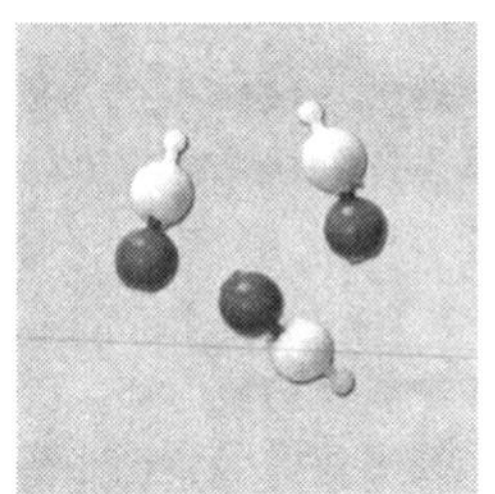

B: Mixture
Mixture of elements; one diatomic (Y_2; Y-Y) and one monatomic (X)

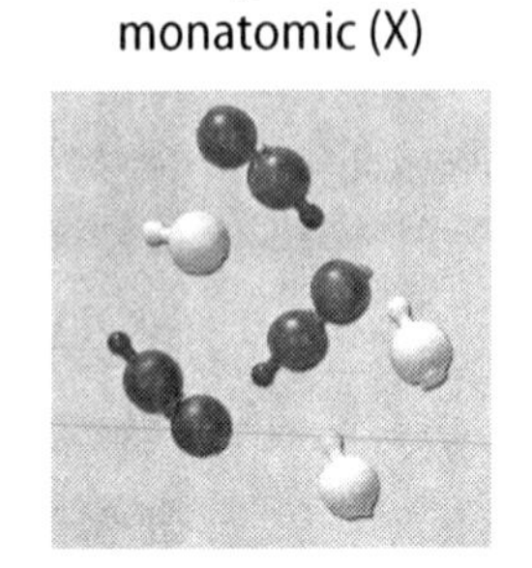

C: Pure Substance
Compound (X_2Y; X-Y-X)

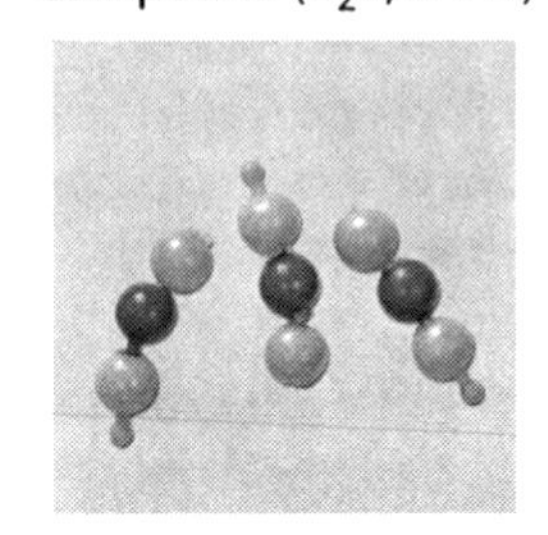

D: Pure Substance
Element (X)

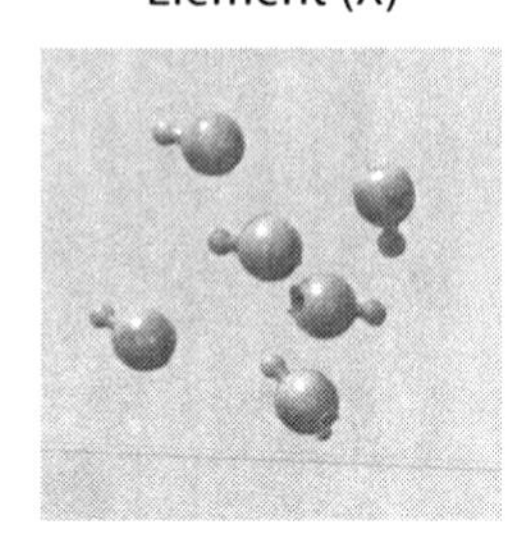

E: Mixture
Compounds-Isomers
(X_2Y_2; X-Y-X-Y and X-X-Y-Y)

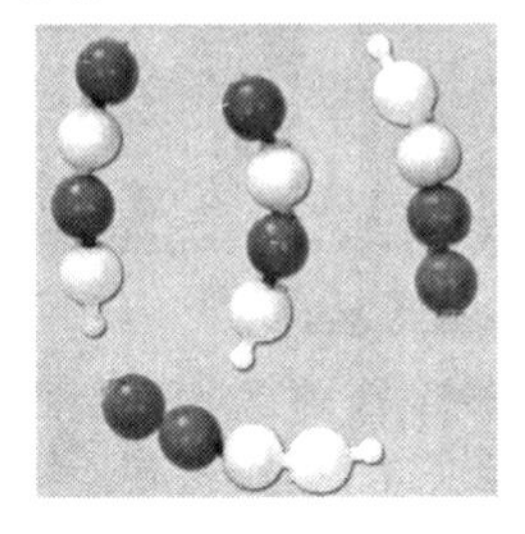

F: Mixture
Mixture of compounds; one with formula (X_2Y; X-Y-X) and other with formula (XY; X-Y)

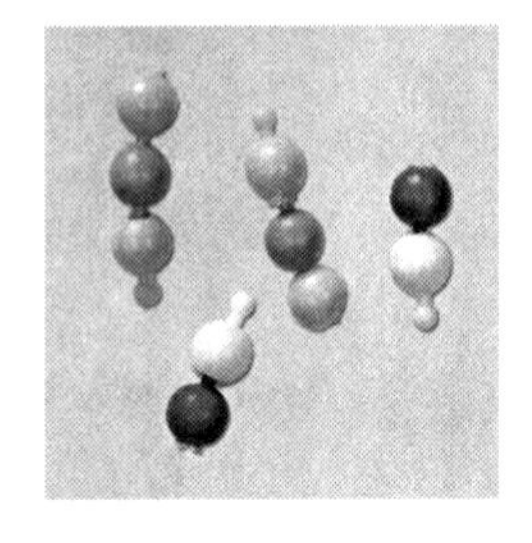

G: Pure Substance
Diatomic element (X_2; X-X)

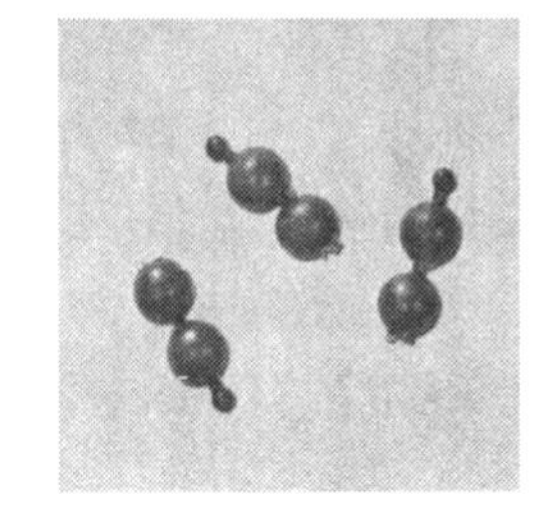

H: Pure Substance
Compound
(X_2Y_2; X-X-Y-Y)

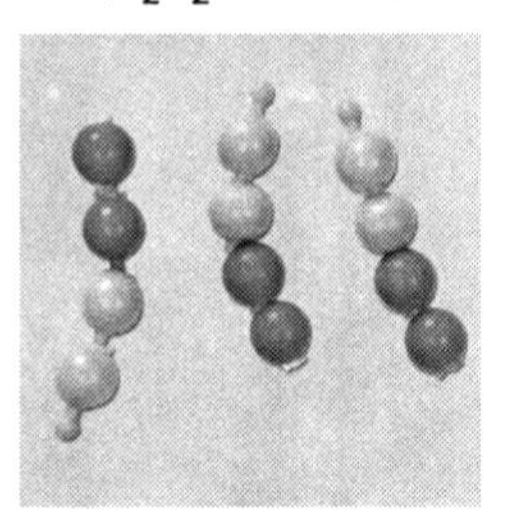

- For Part C, create a bag "P" containing a random collection of unassembled pop-beads.
- For Part C, step 1, you may want to give students more explicit instructions, such as the following: Examine the sample bags one at a time. (Do not remove the contents.) For each bag, first determine whether the contents represent a pure substance or a mixture. Second, if the contents of the bag represent a mixture, determine what the mixture is made of (for example, a mixture of elements and compounds). For the pure substances, determine whether the contents represent an element or a compound. Third, write a chemical formula to represent each of the various "pure substances" in the bag.
- During Part C, step 3, let students grapple with discovering the need to use something other than a molecular formula to represent the two different

arrangements, but refrain from discussing structural formulas until after students complete step 3. After step 3, lead students to understand that they can use structural formulas to show arrangement of atoms in a molecule. After step 4, lead students to understand that bag E contains two isomers.

Explanation

The models used in this activity are intended to represent atoms and covalent compounds, not ionic compounds.

In Part A, students view models of monatomic and diatomic elements, compounds, and a mixture of elements and/or compounds. Atoms are the building blocks of matter and are the smallest unit of matter that cannot be further subdivided by ordinary chemical means. Elements are pure substances. An element is comprised of atoms with identical chemical properties.

As defined by the International Union of Pure and Applied Chemistry (IUPAC), a molecule is "an electrically neutral entity consisting of more than one atom." (Check your textbook—this definition varies depending on the source.) For example, the elements nitrogen and oxygen are found in nature as diatomic molecules (represented by N_2 and O_2, respectively).

A compound consists of two or more different elements. A water molecule, for example, contains two hydrogen atoms bonded to an oxygen atom. Compounds are pure substances. A compound has a definite composition that cannot vary. Compounds can be broken down by chemical reactions into their component elements. The properties of a compound are different than the properties of its constituent elements.

A mixture is a physical combination of two or more different pure substances (elements or compounds). Each kind of matter in a mixture keeps its own chemical identity. Unlike compounds, mixtures can have any composition and the composition can vary in different parts of a mixture. Mixtures can be separated into their components by some physical process. The properties of the components are the same if they are alone or in the mixture.

An important difference between separating compounds into elements and separating mixtures of elements is modeled in Part B. The Velcro in the compounds represents chemical bonding. Using this very simple model, students discover that separating a compound into its elements requires breaking its chemical bonds and, therefore, requires more energy than separating a mixture into its components. The resulting elements have different chemical properties then the compounds they were originally in.

In Part C, students revisit the concepts introduced in Part A using pop-bead models of atoms and molecules. Part C also incorporates the process of writing chemical formulas.

As students discover in Part C, a molecular formula only represents the number(s) of each atom in a molecule. The structural formula is a representation of the molecule that shows how the atoms are arranged. Molecules with the same molecular formula but different structural formulas are called isomers. Isomers have different chemical and physical properties.

Answers to Student Questions

Part C

Step 1

See photo captions in Tips and Instructional Strategies for the answers.

Step 3

a. Yes

b. A possible different arrangement would be XXY.

Step 4

X-Y-X-Y and X-X-Y-Y

Reference

IUPAC. *Compendium of Chemical Technology;* McNaught, A.D., Wilkinson, A., Eds.; Cambridge, England, 1997 (commonly referred to as the "gold book").

Modeling the Behavior of Water

Overview

Engage your students with this fun set of demonstrations that use ceramic magnets to model the polar nature of water molecules. Students will see how one water molecules' attraction for another makes water "sticky" and why water is a good solvent for some substances but not others.

Key Concepts

- hydrogen bonding
- ionic substances
- ionic bonding
- models
- polar and nonpolar substances
- solubility
- water and its properties

National Science Education Standards

Science as Inquiry

Abilities Necessary to Do Scientific Inquiry

- *Students make observations and answer questions about the behavior of magnets that model water molecules, cations, and anions. (5–8, 9–12)*
- *Students learn how magnets can be used to describe and model the behavior of the attractive forces between atoms and molecules. (5–8, 9–12)*

Physical Science

Motions and Forces

- *Students observe how magnets can attract and repel each other while having no effect on certain other materials. (5–8)*

Structure of Atoms

- *Students learn that water is made of molecules, which in turn are made up of hydrogen and oxygen atoms. (9–12)*

Structure and Properties of Matter

- *Students see that the physical properties of water reflect the nature of the interactions among its molecules. (9–12)*
- *Students learn that hydrogen bonding is a type of intermolecular attraction. (9–12)*
- *Students learn that water will dissolve some substances and not other substances as a result of the polar or nonpolar nature of the substances. (9–12)*
- *Students discover how table salt, an ionic compound, is bound together in a crystalline solid. (9–12)*

Part A: Teacher Demonstration

How can water be sticky?

Materials

- magnetic water models
- overhead projector

Procedure

➤ *All magnets must be placed "right side up;" otherwise, the polarity of the magnets will be reversed.*

❶ Place the magnetic water models plastic side down on an overhead projector platform and turn the projector on. Tell the students each piece represents a water molecule. Ask the class to describe the shape of the "molecules." Ask how many different atoms are present in one water molecule and what atoms they are. Which magnet(s) represent which kind of atom(s)?

❷ Slide the large magnet of one water molecule close to the large magnet of another water molecule, as shown in Figure 1. Ask students to describe what happens.

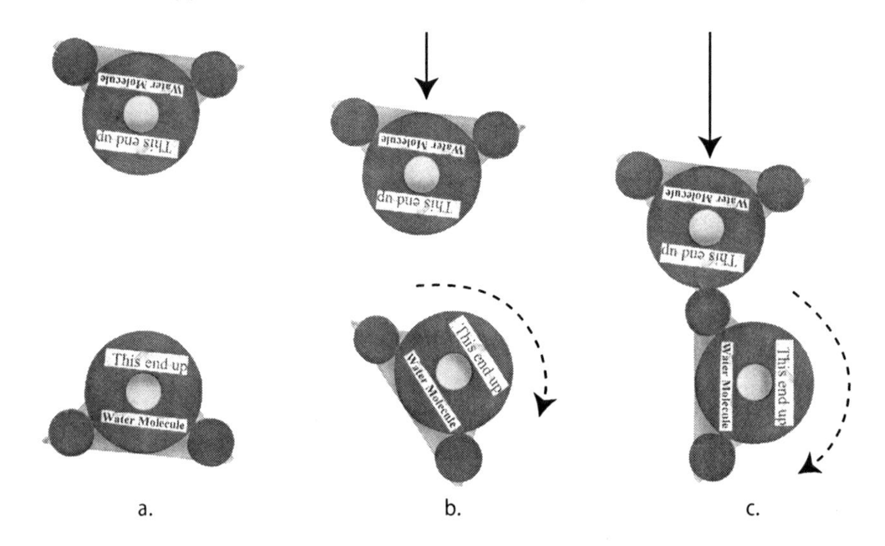

a. Start pushing one water molecule towards another, oxygen sides facing.
b. The molecule you are moving towards begins to rotate.
c. The attraction of one water molecule toward another is modeled.

Figure 1

3. Slide a third water molecule toward the two you have joined together, as shown in Figure 2. Ask students to describe their observations and what conclusions can be made about the nature of oxygen and hydrogen in a water molecule.

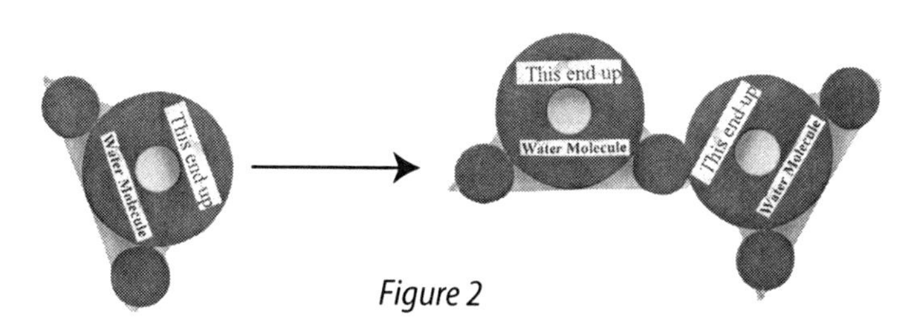

Figure 2

Part B: Teacher Demonstration

Model ionic salts and how they dissolve in water.

Materials

- magnetic water and ion models

➤ *The single donut-shaped ceramic magnets with colored plastic centers represent ions. Magnets with red centers represent positive ions (cations) and those with blue centers represent negative ions (anions).*

- overhead projector

Procedure

 *Be careful; the magnets are strong enough to jump or flip, pinching your fingers as you do the demo.*

1. Spread out the anion and cation models, colored side up, on the overhead. Tell the students that these represent positively and negatively charged particles of matter. Have students note any movement as you slowly move an anion toward another anion, then toward a cation. Repeat with a cation.

2. Ask the students to predict how the particles will arrange themselves when they are moved close together. Push the magnets close together to test their prediction. The positive and negative ions will alternate. This orderly arrangement is an example of how the particles of an ionic compound, such as table salt (NaCl), are arranged in a crystalline solid.

3. Move the oxygen end of a water model toward an anion in the crystalline solid as shown in Figure 3. Have the students note the result. Now move the hydrogen end towards an anion. Repeat this process with a cation. Ask students what their observations tell them about the water molecule.

Figure 3

❹ Hold the water molecule so that its oxygen is attracted to a cation. Pull this water-ion unit away from the other ions. (Hold the other ions to keep them from moving.) Other nearby water molecules will be immediately drawn toward the water molecule and ion and arrange themselves so the ends of the water molecules with charges opposite that on the ion are close to the ion. Each ion becomes surrounded by water molecules as shown in Figure 4. This models how water dissolves ionic compounds.

Figure 4

Part C: Teacher Demonstration

So, what about nonpolar molecules?

Materials

- magnetic water models
- pennies
- overhead projector

Procedure

1. Spread out several pennies on the overhead. Ask the students to speculate on what these pennies represent (nonpolar molecules). Have students note the lack of attraction as you move one of the pennies towards another.

2. Push all of the pennies together. Ask the students to predict what will happen when water molecules are brought next to the nonpolar substance. Try this with a water molecule with the oxygen pointed at the nonpolar substance, then try again with a hydrogen pointed at the nonpolar substance. (See Figure 5.) Ask students to explain what they observed.

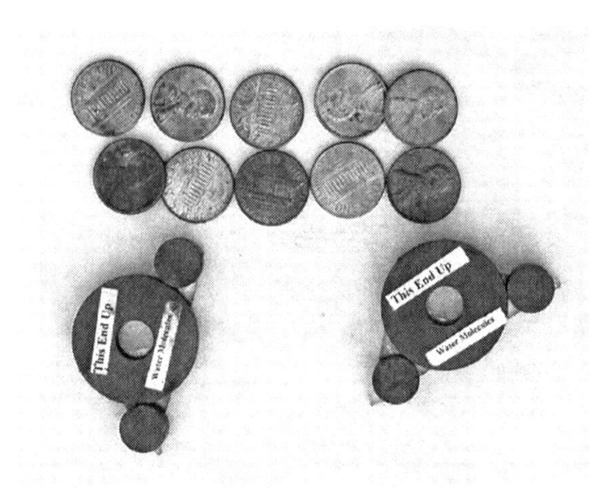

Figure 5

Instructor Notes

- A water molecule model can be made using epoxy to adhere magnets to a triangular piece of Plexiglas™ (for stability). Magnets that represent oxygen must have the same polarity and those representing hydrogen must have the opposite polarity. (You can change the polarity of a round magnet by flipping it over.) An ion model can be made from colored acetate adhered to a donut magnet.
- The magnetic water and ion models do not perfectly reflect the nature of real water molecules. Like all models, the magnets cannot reproduce or replicate all aspects of the real thing. Some of the limitations of the model are as follows:
  - Neither water molecules nor atoms of oxygen or hydrogen are magnetic. Water molecules are attracted to each other as a result of a force between molecules called hydrogen bonding.
  - The charges on ions are electrical in nature, not magnetic.
  - The relative sizes of the atoms in the model are not to scale, and the H-O-H bond angle in the model is not accurate.
  - The magnetic water models could possibly generate orientations that do not reflect commonly accepted representations of hydrogen bonding in water. This can occur because interactions between water molecules are three-dimensional while the models are two-dimensional.
  - The ceramic magnets used in these models were magnetic before they were attached to each other and remain magnetic even after they are glued to the Plexiglas. On the other hand, the polarity of an O-H bond is the result of the chemical transformation that occurs when the molecules of nonpolar hydrogen (H_2) and nonpolar oxygen (O_2) react to form water. The overall polarity of the water molecule is due to the specific geometric arrangement of the atoms within this molecule, which is influenced by nonbonding electron pairs of the oxygen that are not typically shown when drawing the structure of a water molecule.

Explanation

Water is pretty amazing! Not only does it cover most of our planet and make up about 70% of the average human body, it is a magnificent molecule that has some really unusual properties, which are primarily a result of water's polar nature.

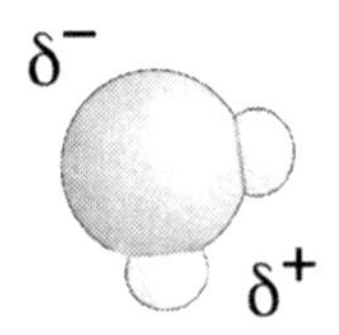

Polar molecules behave as though they have a partial positive charge (δ+) at one end and a partial negative charge (δ-) at the other end. (See figure.) The symbol δ is read "partial," meaning the electrical charge is considerably less than a full positive or negative charge. The attraction between the δ+ of one molecule and the δ- of another molecule results from the electrostatic attraction, not a magnetic attraction as our models do. In water, the difference in electronegativity between a hydrogen and an oxygen is very large, resulting in the polarity of the O-H bonds. This, together with the bent geometry of the molecule and the small size of the hydrogens causes water to have an exceptionally strong attraction to itself. This effect is modeled in Part A when one water molecule magnet is moved towards another.

Part B begins with a representation of an ionic crystalline solid, such as table salt (NaCl), using magnets as models for the cations and anions. Of course, in a real ionic compound, the ions are held together by ionic bonds that result from strong electrostatic attraction between oppositely charged ions. The 1:1 arrangement of the sodium ion (Na^+) and chloride ion (Cl^-) is easily demonstrated in this model, although the model does not reflect the three-dimensional nature of the crystal.

Part B continues with a representation of how ionic solids can dissolve in water. The positive ends of water molecules are attracted to anions (negatively charged), and the negative ends of water molecules are attracted to cations (positively charged). As more and more of the polar water molecules orient themselves around the ions in the outer regions through a process called hydration, the crystal starts to break up into individual cations and anions, each surrounded by a number of water molecules. This shielding prevents the oppositely charged ions from reforming the crystalline solid. The resulting hydrated ions move in clusters throughout the solution that results from this dissolving process.

In Part C, students observe how differently water behaves with nonpolar substances (represented by pennies). Nonpolar molecules are not charged (as the ions were in Part B), nor do they have a dipole as polar molecules (like water) do. As a result, nonpolar molecules generally do not dissolve in water.

Reference

Davies, W.G. Magnetic Models of Ions and Water Molecules for Overhead Projection. *J. Chem. Educ.* **1991,** *68*, 245–246.

Magic Sand

Overview

This engaging activity focuses on the behavior of Magic Sand™, which is coated with a waterproof (hydrophobic) material. In Part A, students explore and compare the behavior of water drops on spoons coated with Magic Sand and spoons coated with regular sand. Part B is a teacher demonstration of Magic Sand's behavior when submerged in a polar substance (water). Part C shows the contrasting behavior of Magic Sand in a nonpolar substance (hexane, vegetable oil, or mineral oil).

Key Concepts

- experimental design
- hydrophilic substances
- hydrophobic substances
- polar and nonpolar substances
- water and its properties

National Science Education Standards

Science as Inquiry

Abilities Necessary to Do Scientific Inquiry

- *Students make and record observations and ask questions about the behavior of water when in contact with a hydrophobic substance. (5–8, 9–12)*
- *Students pose testable questions and design experiments to answer these questions. (5–8, 9–12)*

Physical Science

Properties and Changes of Properties in Matter

- *Students observe that the abilities to repel or attract water are characteristic properties of matter. Students also observe that sand can be separated from water using physical properties of the substances in a sand-water mixture. (5–8)*

Structure and Properties of Matter

- *The atoms and resulting structures of molecules determine whether a substance is polar or nonpolar and likewise whether the substance is attracted to or repelled by other substances. (9–12)*

Science and Technology

Abilities to Do Technological Design

- *Students evaluate possible uses of hydrophobic materials in products. (5–8, 9–12)*

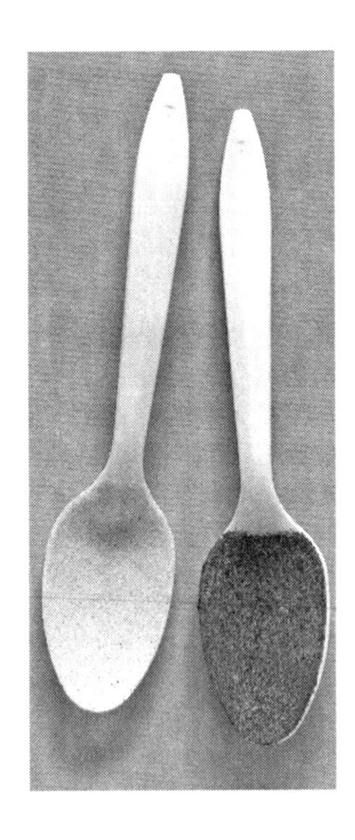

Part A: Student Exploration

Sandy spoons—How different are they?

Materials

- set of sand-covered spoons
- water

Procedure

⚠ *Avoid rubbing the material coating the spoons excessively as it may rub off. Avoid ingesting any of the materials used in this activity.*

1. Observe both of the spoons. Make a list of similarities and differences.
2. Place a small amount of water in each spoon. Explore how water acts in each spoon. *Suggest a reason for the differences you observe.*
3. Working with a partner, experiment with different ways of transferring the water back and forth between the spoons. Compare and contrast transferring the water between spoons coated with the same and different substances.

Part B: Teacher Demonstration

Compare beach sand with Magic Sand.

Materials

- Magic Sand
- regular sand
- 2 clear containers
- water
- 2 plastic spoons
- paper towels

Procedure

1. Partially fill one of the clear containers with water. Pour a few spoonfuls of regular sand into the water. Show this around for the class to observe. Use a spoon to lift some of the sand out of the container and then allow this sand to spill back into the container. Challenge the students to explain what they observed.
2. In front of the class, decant as much of the water from the container as you can and then pour the sand onto a dry paper towel. Ask the class to discuss their observations and possible explanations.
3. Repeat step 1 with Magic Sand, asking students to compare their observations to those previously made.
4. Sprinkle some Magic Sand on the surface of the water so that a layer of the sand floats on the water (due to surface tension). Instruct the class to carefully observe while a student volunteer slowly pushes his or her fingertip downward through the floating sand and into the water. Stop the student before he or she pushes so deeply that water rushes over the top of the sand-coated finger. Ask the class what they observe. Have the volunteer remove his or her finger from the sand and describe what the finger feels like.
5. Instruct the class to observe as you decant the water and pour the Magic Sand onto a dry paper towel as in step 2. Discuss observations and possible explanations.

Part C: Teacher Demonstration

What is the clear, colorless liquid?

Materials

- Magic Sand
- clear glass container
- nonpolar liquid such as hexane, vegetable oil, or mineral oil

➤ *Vegetable oil or mineral oil is apt to render the Magic Sand essentially useless for other applications.*

Procedure

⚠ *Hexane is a volatile respiratory irritant and should be used in a well-ventilated area. Goggles are required. We recommend that hexane be used only by the instructor. Dispose of according to local ordinances.*

1. Challenge your students to determine the type of solvent that is being used in this part of the demonstration.
2. Repeat step 1 in Part B using Magic Sand, a clear glass container, and hexane or oil. Tell the students to compare the observations made here to those made in Part B. Ask students if this liquid is water and have them justify their responses.
3. Ask students to predict what will happen if you add water to this mixture. Do this and challenge them to explain their observations.
4. Decant the liquids from the Magic Sand. Dispose of the solvents according to local ordinances. Pour the Magic Sand onto a dry paper towel. Pat to dry the sand with another paper towel. Return the Magic Sand to the beaker and add water as you did in Part B. Challenge the students to explain their observations.

Instructor Notes

Tips and Instructional Strategies

- To make the spoon set for Part A, spray the bowls of two plastic spoons with adhesive (such as 3M™ Super 77™ Multipurpose Adhesive, available at craft stores) and sprinkle one spoon with Magic Sand and the other with beach sand. Work in a fume hood or well-ventilated area. Allow to dry overnight.
- Magic Sand should not be used in fish tanks as it may kill the fish.
- Both types of sand can be saved for reuse, although as noted in Part C, oils may ruin the Magic Sand.
- Students may be interested to learn that, in addition to being an interesting and educational novelty toy, Magic Sand has practical applications based on its hydrophobic properties. For example, in the arctic, utility companies can bury junction boxes in Magic Sand. Since Magic Sand is hydrophobic, it doesn't absorb water like soil does. Therefore, Magic Sand remains dry and loose in extremely cold temperatures, so that the junction box can be serviced even when the ground is frozen.
- In Part A, after students have explored the two spoons, tell students that one of the spoons is coated with regular beach sand. Ask them to identify which one that is. Tell them the colored sand is regular sand that has a special coating on it. Ask them to classify this coating as hydrophobic or hydrophilic. (For older students, a discussion of nonpolar and polar molecules is also appropriate.)
- In Part A, step 3, you can encourage students to make a game of tossing water back and forth with the spoons and point out that because of hydrogen bonding between water molecules, the water stays together as it travels through the air.
- After students have observed the interaction of Magic Sand and water in Part B, you may want to pour some dry Magic Sand onto a paper towel to show students that without water, the Magic Sand does not seem so "magical." The water allows the Magic Sand to exhibit the unique hydrophobic properties that cause it to form unusual shapes.
- Challenge students to pose testable questions based on what they've done and learned in the lesson. They can design experiments and collect evidence to answer these questions, formulate claims about their findings, and, if time allows, present and defend their claims with their group or class. While we strongly recommend that students develop their own testable questions, you may need to seed the discussion with possible questions such as these: How would our results differ if we tried different liquids in Part A, step 3? (Remember that oils may ruin the Magic Sand.) How would the Magic Sand behave if it were added to a water-detergent solution? Does the type of adhesive used to make the sand-covered spoons affect the results?

Explanation

Beach sand gets wet due to the adhesion of water on its surface. On the other hand, Magic Sand remains dry in water, exhibiting a hydrophobic effect due to its coating. (Micrographs of the surface of the Magic Sand show that this phenomena is much more complicated than discussed in this document. For a discussion of free energy, surface tension, and more, see the *Journal of Chemical Education* article by Vitz listed in the References section.) The Magic Sand coating creates a hydrophobic surface that does not attract water. In Part A, students explore the interactions of both types of sand with water, discovering the hydrophobic nature of the Magic Sand on their own.

In Part B, students observe the hydrophobic effect with a quantity of Magic Sand, which remains dry even when totally submerged in water. When a student pushes down on a layer of floating sand, sand surrounds the finger and water is repelled. A visual phenomenon, internal reflection, may be observed with the submerged Magic Sand. As the Magic Sand falls through the water, it gathers trapped air to its surface, forming a sort of bubble around the sand. Light reflects off the sand and some of the light reflects off the bubble back toward the sand. This internal reflection is seen as a "silvery" appearance on the submerged sand.

As evidenced in Part C, Magic Sand exhibits its hydrophobic nature only when it is in water. With either nonpolar hexane or oil, the Magic Sand does not mold into structures that hold their shape in the liquid.

With hexane, a nonpolar short chain hydrocarbon, some of the dye coating of the Magic Sand is dissolved, but not enough to prevents its hydrophobic characteristics from exhibiting themselves once water is introduced into the system. Another interesting phenomenon is also observed: Due to the immiscibility of the hexane and water and the density differences of the hexane, water, and Magic Sand, a three-phase system results. Dye-colored hexane occupies the top layer, colorless water the next, and solid Magic Sand the bottom layer.

With vegetable or mineral oil, results are somewhat different. The oil does not dissolve the dye so the oil does not take on the dye's color. While a three-phase system will result upon the addition of water (due to density differences), the oil layer may be less visible if you are using mineral oil, which is colorless.

Answers to Student Questions

Part A

Step 2

One type of sand attracts water and the other repels water. (In other words, untreated sand is "wetted" by water and Magic Sand is not.)

References

Goldsmith, R.H. Illustrating the Properties of Magic Sand, *J. Chem. Educ.* **2000**, *77*, 41.

Sarquis, J.; Sarquis, M.; Williams, J. *Teaching Chemistry with TOYS;* Terrific Science Press: Middletown, OH, 1995.

Vitz, E. Magic Sand: Modeling the Hydrophobic Effect and Reversed-Phase Liquid Chromatography, *J. Chem. Educ.* **1990,** *67* (6), 512.

Colorful Lather Printing

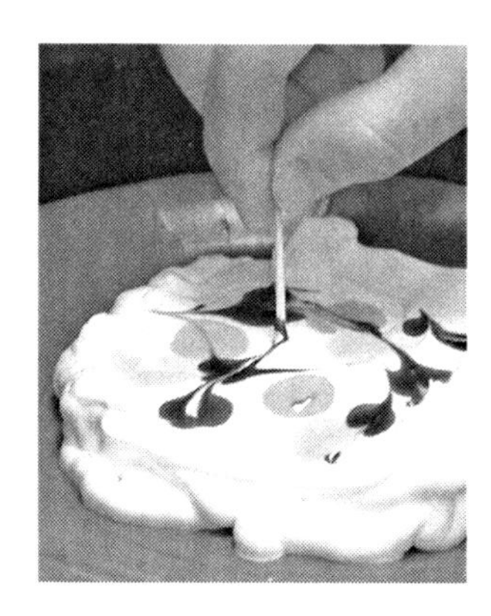

Overview

In this activity, students explore the chemistry of soap while using shaving cream, a common soap lather, to create beautiful colored patterns and capture the patterns on paper.

Key Concepts

- hydrophobic substances
- ionic compounds
- polar and nonpolar substances
- soaps and detergents
- surface tension

National Science Education Standards

Science as Inquiry

Abilities Necessary to Do Scientific Inquiry

- *Students use logic and evidence to formulate explanations about what they observe of the interactions among shaving cream, paper, food color, water, and other solutions. (5–8, 9–12)*

Physical Science

Properties and Changes of Properties in Matter

- *Students observe that food color spreads on paper, dissolves in water, and spreads less on shaving cream than in water. They also learn about the properties of hydrophobic substances and polarity. (5–8)*

Structure and Properties of Matter

- *Students observe and learn about the ways that different substances interact with each other due to the nature of the molecules. (9–12)*
- *Students conclude that the polarity of substances helps determine certain interactions between substances. (9–12)*
- *Students observe the effects of soap on the surface tension of water. (9–12)*

Part A: Student Exploration

Create art while observing the properties of shaving cream, a soap lather.

Materials

- food colors in dropper bottles
- 3–4 pieces of a white, nonglossy paper

➤ *A sturdy paper, such as used in index cards or art paper, works well.*

- small clear cup
- water
- aerosol shaving cream (standard white type)
- paper plate
- cooking spatula or craft stick
- toothpicks
- dropper or straw
- paper towels for cleanup

Procedure

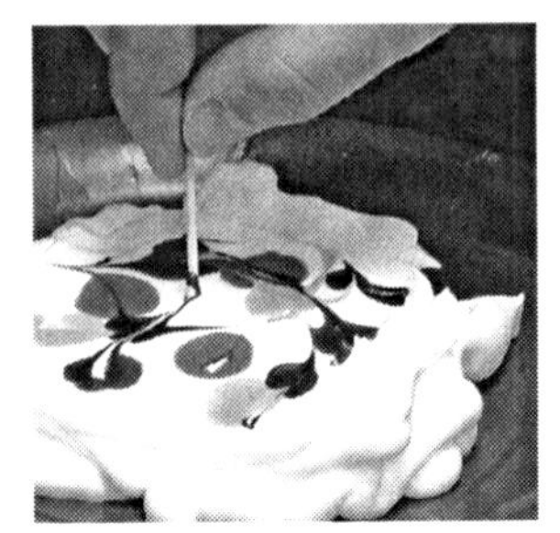

1. Place one drop of food color onto a sheet of paper. *Observe the interaction of the drop and the paper.* Notice the extent to which the color spreads on the paper. *Describe and record what you observe.*

2. Fill a cup about half-full with room-temperature water. Without stirring, add one drop of food color to the water. Notice the extent to which the color spreads through the water. *Observe the interaction of the drop on the surface and in the water. Describe and record what you observe.*

3. Dispense a pile of shaving cream roughly the size of your fist onto the paper plate. Using a spatula or craft stick, shape the shaving cream so that the top surface is nearly flat and the area of this surface is slightly larger than the paper you will be using. Apply drops of several different colors of food color to different locations on the shaving cream. (A total of 6–8 drops works well.) Notice the degree to which the color spreads through the shaving cream. *Observe the interaction of the drops with the shaving cream. Describe and record what you observe.*

4. Drag a toothpick through the colored drops on the shaving cream to create patterns with the color. Drag colored shaving cream into uncolored areas or uncolored shaving cream into colored areas. Try making different patterns. Straight lines, curved lines, parallel lines, and spirals will all produce different effects.

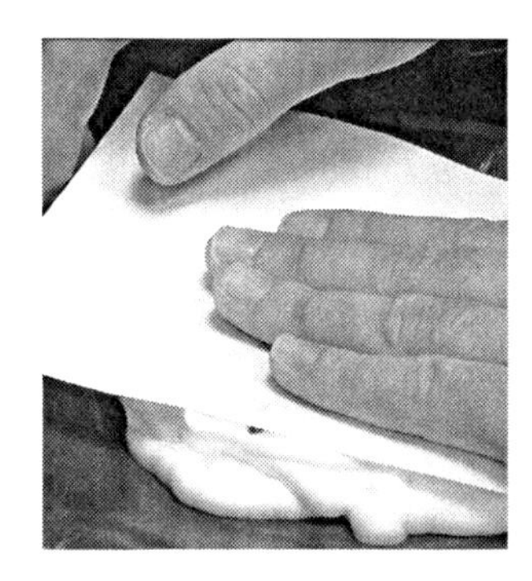

5. Gently press the paper onto the surface of the patterned shaving cream. Lift the paper off the shaving cream. Scrape off the excess shaving cream close to the paper using the spatula (or craft stick) and return the excess shaving cream to the original pile. *What do you observe on your paper? Why do you think this happens?*
6. Using the spatula (or craft stick), mix the pile of colored shaving cream until it is one uniform color. If the color is very pale, mix in a few more drops of food color.
7. Apply a single drop of water to the surface of the tinted shaving cream and observe. Add a few more drops of water at different places on the surface for design purposes. *What do you observe?*
8. Make a print from your new mixture following the method you used in step 5. Compare your print with others in your class.

Part B: Student Exploration

Continue to explore the surface effect of detergent and discover the effect of tinting the shaving cream with different colorants.

Materials

- aerosol shaving cream (standard white type)
- concentrated copper sulfate ($CuSO_4$) or cobalt chloride ($CoCl_2$) solution

➤ *The exact concentration of the solution is not important; the solution is used to produce a vibrant color in step 3.*

- food colors in dropper bottles
- clear cups
- light (in color) corn syrup
- ice
- liquid detergent, such as Dawn® or Joy®
- toothpicks

Procedure

1. Pour about 2.5 cm (1 inch) light corn syrup into a clear cup. Add 3 or 4 drops of food color onto the surface of the corn syrup. *What do you observe?* Dip the end of a toothpick into liquid detergent. Wipe any excess detergent off the toothpick. Touch the toothpick to the surface of the corn syrup in between the drops of food color. *What do you observe? Why do you think this happens? How is the behavior similar to what you observed on the tinted shaving cream?*

2. Pour about 2.5 cm (1 inch) water into a clear cup. Add ice to the water and stir. After a few minutes, remove the ice that has not melted. Drop 3 or 4 drops of food color in various locations on the surface of the cold water. *What do you observe?* Touch a new toothpick containing detergent on the surface of the water between the drops of food color. *What do you observe? How is this behavior related to that seen when you added water to the surface of the tinted shaving cream?*

⚠ *Cobalt chloride ($CoCl_2$) and copper sulfate ($CuSO_4$) are both harmful if swallowed. In addition, both are skin and respiratory irritants. Goggles are required. Do not ingest; use with adequate ventilation; and wash well with water in case of accidental skin contact. Dispose of according to local regulations.*

3. Add either concentrated copper sulfate ($CuSO_4$) or cobalt chloride ($CoCl_2$) solution to a fresh pile of shaving cream and mix until an intense color is achieved. Now try applying a drop of water to the surface of the tinted shaving cream. *What do you observe? How does the behavior compare with what you observed in Part A? How are $CuSO_4$ (aq) and $CoCl_2$ (aq) different than food color?* Look up the formulas for common FD&C food colors on the Internet. Compare these to the formula of $CuSO_4$ or $CoCl_2$.

Instructor Notes

Tips and Instructional Strategies

- Food color can stain furniture. If necessary, protect work surfaces with newspaper or plastic sheeting.
- Part A of this activity is appropriate for all levels. Part B, a more sophisticated investigation appropriate for advanced high school or college, challenges students to consider the effects of size and polarizability of ions and molecules in solution.
- Emphasize to students that only one drop of food color should be used in Part A, steps 1 and 2.

- Students should avoid piling too much shaving cream on the paper plate in Part A, step 3, to minimize the mess when shaving cream is scraped off the paper in step 5.
- Be sure students put only drop-size amounts of food color on the shaving cream surfaces. Close monitoring of this step may be needed, depending on the abilities of the students.
- If needed, explain to students that shaving cream is a lather, similar to a foam, and a foam is a colloid (a gas trapped within a liquid). Have students do research to answer the following questions: What other common products are foam or lather colloids? How are colloids, in general, different from solutions? What do solutions and colloids have in common?
- In Part A, step 7, ask students to try to explain their observations. While they're not likely to have a full grasp of the chemistry involved, they should be able to surmise that the soap on the surface of the water is somehow "pushing" the color away.
- In Part B, be sure that students understand how the investigations are related to what they observed in Part A, which entailed a surface effect of adding water to tinted shaving cream. In steps 1 and 2 of Part B, they are adding soap or detergent to simpler systems, including corn syrup (sugar water) and ice water. In step 3, they are using a different colorant to explore the effect of the size of dissolved ions on the phenomena.
- In Part B, step 3, you may also want to have students try colored indicator solutions, such as methylene blue, or tempera or other paints.

Explanation

In this activity, shaving cream (a soap lather) is used as a support for the food color marbling patterns transferred to white paper. Students marble paper with shaving cream and food color as they learn about the concepts of polarity and hydrophobicity. While this is a familiar activity to many educators, a new twist is added here—exploring how the colored shaving cream mixture behaves when a drop of water is added. You and your students are sure to be surprised by the results. This new dimension helps students to further refine their mental pictures of the science of the system as well as the nature of soap, surfactants, solutions, ions, colloids, and diffusion.

During Part A, step 1, the food color spreads across the paper due to wetting, surface tension lowering, and capillary action. After observing how food color spreads in water (step 2), students discover in step 3 that color spreads in shaving cream less than in water because shaving cream is a lather. Shaving cream is a mixture of a liquid (soap suspended in water), tiny bubbles of the propellant gas (butane), and solid soap, which makes it a lather. The viscosity of the lather retards diffusion of the food color until the design is transferred onto the water-absorbent paper (step 5).

In Part A, step 6, the food color is mixed into the shaving cream to create a uniformly colored mixture. When water is dropped onto the surface of this mixture, a white spot immediately forms. This spot results from the lowering of the surface tension in the water at the point the drop contacts the soap in the shaving cream. The food color originally present in the area is repelled as the surface tension is lowered.

Food colors are large organic species. Figure 1 shows the chemical structures for FD&C Red No. 40 and Blue No. 1. The water solubility of these FD&C colorants is due to the presence of sodium carboxylate and sodium sulfonate groups in their structures. These species are both large and polarizable. (According to Brown, LeMay, and Bursten, "Polarizability is the ease with which the electron cloud of an atom or a molecule is distorted by an outside influence, thereby inducing a dipole moment.") It is the size and polarizability of the FD&C food colors that make them surface acting agents.

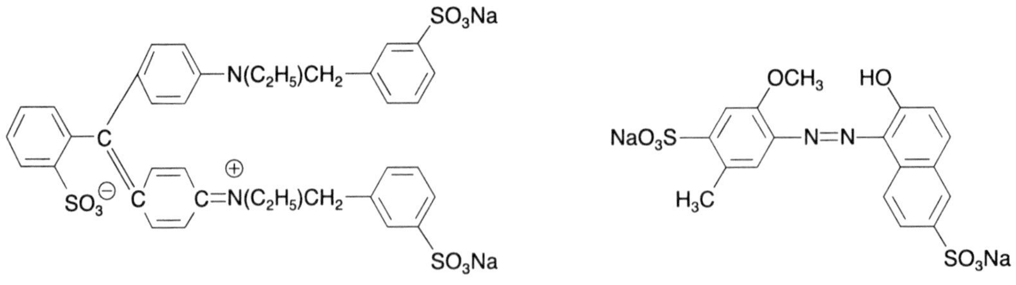

FD&C Blue No. 1

FD&C Red No. 40

Figure 1: FD&C colorants

In Part B, students explore the effect of changing the ions used to color the shaving cream. No change occurs when water is added to shaving cream colored with either Cu^{2+} or Co^{2+}. This is because the smaller Cu^{2+} and Co^{2+} are not very polarizable and are not surface active.

The surface activity of the food color is further demonstrated on corn syrup, a viscous sugar-water solution. A drop of food color on the surface of the syrup is repelled when a toothpick containing detergent is applied to the surface. With ice water, most of the food color diffuses into the water; however, the small amount of food color that remains on the surface is also repelled when the detergent-containing toothpick is applied to the surface.

Answers to Student Questions

Part A

Step 1

The color is absorbed by the paper but spreads very little.

Steps 2 and 3

The color spreads a little more in the shaving cream than on the paper, but spreads far less than the rapid spreading (or diffusion) observed in water.

Step 5

The marble pattern is transferred to the paper because the paper absorbs the food color.

Step 7

A white spot forms almost immediately where the drop of water falls on the tinted shaving cream.

Part B

Step 1

a. *A thin film of food color spreads out on the surface of the syrup, and the rest of the drop of food color sinks into the syrup.*
b. *When the detergent is applied, the thin film of food color on top of the syrup is repelled, creating a star-like, colorless (clear) surface pattern on the thick syrup. Since the syrup is viscous, the pattern forms slowly.*
c. *The colorless pattern that forms on the syrup's surface when the detergent is added is similar to the colorless region that forms when the water drop is dropped onto the surface of the tinted shaving cream.*

Step 2

a. *Most of the food color sinks and disperses in the cold water, but a very thin film spreads out on the water similar to that observed with the syrup. The food color film on the cold water is much thinner than that observed on the surface of the syrup and the color disperses more slowly than in room-temperature water.*
b. *Applying detergent to the thin food color film causes a star-like, colorless pattern to quickly form on the surface. The colored film quickly moves away from the point on the surface at which detergent or soap was added.*
c. *In both cases, food color is repelled from the surface as the detergent and water come into contact.*

Step 3

When a water drop is added to the $CuSO_4$ *or* $CoCl_2$ *tinted shaving cream, the blue or pink color persists. A white spot does not form on the shaving cream surface as it did with food color. The food colors are large ions containing many atoms, while the* Cu^{2+} *and* Co^{2+} *cations (responsible for the color in the* $CuSO_4$ *and* $CoCl_2$ *solutions, respectively) are monatomic.*

References

Brown, T.; LeMay, H.; Bursten, B. *Chemistry: The Central Science,* 11th ed.; Prentice Hall: Upper Saddle River, NJ, 2008.

Hershberger, S.; Nance, M.; Sarquis, M.; Hogue, L. Colorful Lather Printing, *J. Chem. Educ.* **2007**, *84,* 608A.

Making and Breaking Emulsions

Overview

In this lesson, students investigate an oil-water system to understand why separation of the two phases occurs and then use common household ingredients to make and break emulsions.

Key Concepts

- density
- emulsions
- experimental design
- polar and nonpolar substances
- soaps and detergents
- solubility

National Science Education Standards

Science as Inquiry

Abilities Necessary to Do Scientific Inquiry

- *Based on their observations of the solubility of different liquids, students develop descriptions and explanations of why some liquids are soluble or insoluble in other liquids. (5–8)*
- *Students make observations and develop descriptions about how the oil and water mix when an emulsifying agent is added. (5–8)*
- *Students use logic and evidence to formulate explanations about density, polar and nonpolar covalent substances, and emulsions. (9–12)*
- *Students determine the polarities of some everyday liquids based on how these liquids behave when added to an oil and water layered mixture. (9–12)*
- *Students use logic and evidence to formulate explanations about how the emulsion in hand cream is broken and then repaired. (9–12)*
- *Students design and conduct an investigation with other emulsions and emulsifying agents to determine if the sequence of adding a broken emulsion and an emulsifying agent is important. (9–12)*

Physical Science

Properties and Changes of Properties in Matter

- *Students observe the layers in an oil-and-water mixture and conclude that water and oil are not alike and, therefore, do not dissolve in one another. (5–8)*
- *Students compare and discover solubilities by measuring the layers in an oil-and-water mixture and then measuring the layers again after additions of different additives. (5–8)*

- *Students observe that some substances are soluble in each other and some are not. They conclude that an added substance, an emulsifying agent, can prevent the separation of two liquids that are normally insoluble. (5–8)*

Structure and Properties of Matter

- *Students observing the layers in an oil-and-water mixture conclude that water (a polar covalent substance) and oil (a nonpolar covalent substance) are not alike and therefore do not dissolve in one another. (9–12)*
- *Students discover that adding liquid soap or detergent to a layered oil-and-water mixture reduces the height of the oil layer because an emulsion forms when some of the oil becomes suspended in the water. (9–12).*
- *Students observe and conclude that an emulsion is one type of colloid in which one liquid is dispersed in another immiscible liquid. (9–12)*
- *Students investigate and discover that repairing a "broken" emulsion requires adding and mixing small portions of the broken emulsion to the emulsifying agent. (9–12)*

Part A: Student Exploration

Why can't some liquids just get along?

Materials

- tape that can be written on
- pen
- ruler
- small, clear, colorless container with lid
- equal amounts of vegetable oil and water (enough to half-fill the container)
- liquid soap or detergent
- at least 3 of the following additives:
  - tincture of iodine
  - food color
  - mineral oil
  - baby oil
  - rubbing alcohol
  - soft drink

Procedure

1. Place a strip of tape on the container as shown in Figure 1. Add the oil to the container and then the water. Use a pen to mark a line on the tape at the top of each layer. Measure and record the height of each layer. *What observation(s) allow(s) you to determine which substance constitutes the top layer and which one the bottom? What general property of substances is responsible for this positioning? Explain.*

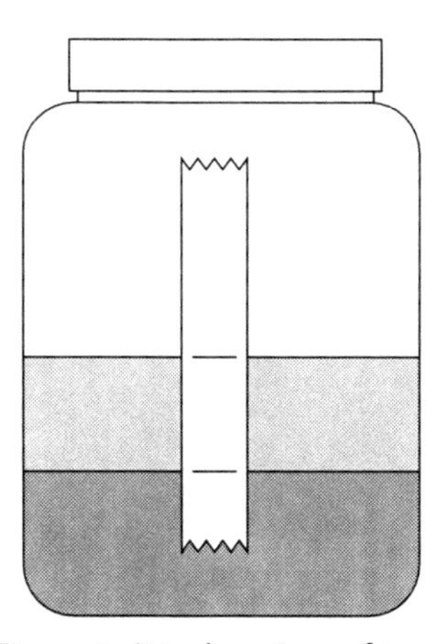

Figure 1: Attach a piece of tape to the outside of the container.

2. Add a little more water to the container, then remeasure the height of each layer. *How can you tell if any of the added water dissolved in the oil? Explain the reason for your observations in terms of polarity.*

3. Add just enough of one of the additives to your oil-and-water mixture to enable you to determine and record its position in the container. Also record the height of each layer and any other significant observations about the mixture. Shake the container well for 15 seconds. Note how long the mixture takes to separate and any changes once the layers have again separated.

4. Repeat step 3 with at least two more of the additives (added one at a time). *What do your observations tell you about the polarity of the liquids you tried?*

5. Add 20–30 drops of liquid soap or detergent to your mixture from step 4. Shake the container well for 15 seconds and allow it to stand for several minutes. Record the height of each layer and any other significant observations about the mixture. *Compare the time it took the mixture to separate into layers with the time it took in step 3. Explain your observations in terms of the nature of soap and detergent molecules.*

6. Investigate two types of Italian salad dressing: creamy style and vinegar-and-oil style. Research to find recipes and read labels from commercial products of both types. *What differences in the key ingredients do you see? Apply what you've learned in this activity to explain the role of a key ingredient in the creamy style that is not found in the other type.*

Part B: Student Exploration

Can you fix a "broken" emulsion?

Materials

- Vaseline® Intensive Care® (preferred) or other hand cream
- water
- 80 mL (⅓ cup) vegetable or mineral oil
- several small plastic cups or containers
- teaspoon measure
- cup measure or graduated cylinder
- liquid dishwashing detergent
- clean towel or tissues
- other materials as needed for the student-designed experiment

Procedure

1. Examine the appearance and feel of the hand cream. Read the label. *What are the key ingredients?*
2. Place about 5 mL (1 teaspoon) hand cream into a small cup. Add about 80 mL (⅓ cup) water and stir for several minutes. Use a dry finger to lightly touch the top of the liquid mixture. *Record your observation(s) of sight and touch. Explain what has happened to the components of the hand cream. What does this tell you about the amount of water that can be "held" by the hand cream?*
3. Pour about 10 mL (2 teaspoons) liquid dishwashing detergent into a clean cup. Stir in about 2 mL (½ teaspoon) of the mixture from step 2. Continue adding a little at a time until all the mixture from step 2 has been stirred in. *Compare your observations to those of step 2. (Neglect any suds.) What can you say has happened to the components of the hand cream?*
4. Pour about 5 mL (1 teaspoon) hand cream into a clean cup. Add about 80 mL (⅓ cup) vegetable or mineral oil and stir periodically for several minutes. Use a dry finger to lightly touch the top of the liquid mixture. *Record your observation(s) of sight and touch, including a comparison to those of step 1. Explain what has happened to the components of the hand cream. What does this tell you about the amount of oil that can be "held" by the hand cream?*
5. Pour about 10 mL (2 teaspoons) liquid dishwashing detergent into a clean cup. Stir in about 2 mL (½ teaspoon) of the mixture from step 4. Continue adding a little at a time until all the mixture has been stirred into the detergent. *Compare your observations to those of step 4. (Neglect any suds.) What can you say has happened to the components of the hand cream?*

6. Design an experiment to determine whether the sequence of adding a broken emulsion and the dishwashing detergent makes a difference. In other words: does adding the broken emulsion to the dishwashing detergent have a different effect from adding the dishwashing detergent to the broken emulsion?

7. Design an experiment to determine the effect of replacing the liquid dishwashing detergent with beaten egg yolk or liquid soap.

Instructor Notes

Tips and Instructional Strategies

- As an extension to Part A, challenge students to design and conduct experiments to determine whether shaking time or vigor affects the amount of time the mixtures take to separate in steps 3 and 5.
- After Part B, step 1, discuss with students what an emulsion is, using hand cream as an example. Point out that they are going to explore how to take an emulsion apart ("break" it) and how to re-emulsify or repair an emulsion once it is "broken."
- If students need help designing an experiment in Part B, step 6, you can lead them to prepare a broken emulsion as in step 2, then add small portions of liquid dishwashing detergent until 2 teaspoons have been added.
- As an extension to Part B, you may want to divide students into groups and have each group do a web search to determine how other common emulsions are made. Groups might choose mayonnaise, French dressing, hollandaise sauce, hair conditioner, or another emulsion. Have each group explain how their selected emulsion could be broken and what might be added to re-emulsify the material.
- Challenge students to explain why mayonnaise is classified as an emulsion. They can also be asked to describe the "cure" for a "broken" emulsion and how it works.
- Challenge students to design an experiment to determine the relative amounts of water in various brands of margarine and "light" margarine.

Explanation

In Part A, a mixture of oil and water is used to characterize the polarity of several other substances by applying the principle of "like-dissolves-like." The "like" refers to the polarity of the substances in a mixture. Polar substances tend to dissolve other polar substances, such as water. Nonpolar substances dissolve other nonpolar substances, such as oil. That is to say, if the substance dissolves in water, it is polar, and if it dissolves in oil, it is nonpolar. For example, most food colors are water-soluble; thus, they are polar and do not dissolve in oil. Rather, the polar food color sinks through the oil layer and dissolves in the water layer. Baby and mineral oils are nonpolar so they will dissolve in the vegetable oil layer. In this activity, two clues can be used to determine which layer the test substance dissolves in. If the test substance is colored, it will change the color of the liquid it dissolves in. If the test substance is colorless, a different clue is needed: we look at the volume change that results (which layer increased in height).

Some mixtures contain both polar and nonpolar substances. Tincture of iodine is a mixture of iodine (I_2, a nonpolar molecule) in ethyl alcohol (CH_3CH_2OH, a molecule with both polar and nonpolar ends). When added to the test mixture, most of the iodine preferentially dissolves in the oil, giving the layer a purple color. A little of the iodine will also dissolve in the water, giving it a brownish color. While ethyl alcohol is soluble in all proportions in water (miscible), a little will also dissolve in the oil.

An emulsion is a type of colloid in which two normally insoluble liquids are mixed such that one liquid becomes dispersed in the other liquid, resulting in a stable mixture. Emulsification prevents the normally insoluble substances from separating, often by adding an emulsifying agent. Many household products and foods are emulsions, including milk, margarine, and mayonnaise.

Soaps and detergents are common emulsifying agents. Their ionic head tends to be soluble in water and the nonpolar tail is soluble in oils. (See Figure 2.)

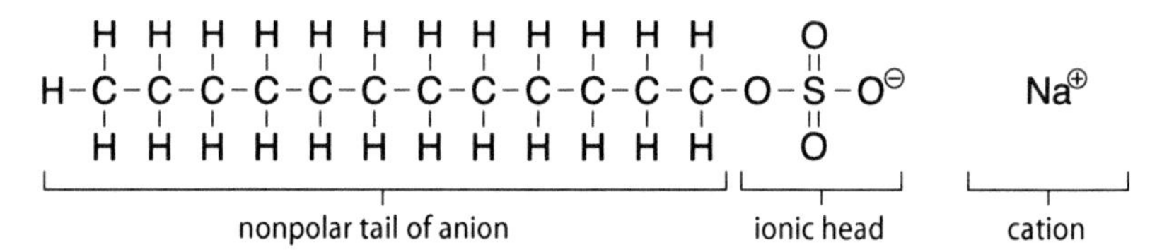

Figure 2: The structure of a detergent

With the addition of detergent, the oil and water remain dispersed in one another for a longer period of time because of the formation of an emulsion. The oil droplets remain small and suspended because the nonpolar tail of the soap is dissolved in them with the ionic heads on the surface, forming a micelle as shown in Figure 3. Since the ionic head has a negative charge, it keeps the oil droplets dispersed since like charges repel. This is why soaps and detergents are used to wash clothes. They suspend nonpolar grease, oil, and fat particles in the water, thus removing them from the clothes.

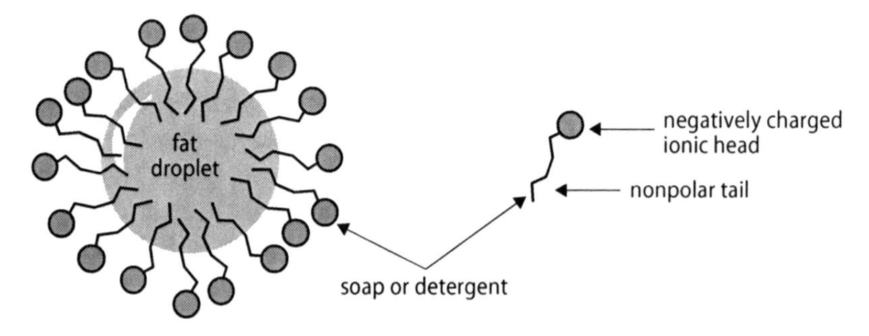

Figure 3: Micelle formation

Gourmet cooks and connoisseurs of fine sauces have long recognized the mark of a classy sauce is one that is smooth and rich, with lots of oil or butter "held" in the water-based mixture. Egg yolk is often the key ingredient. The oil-holding capacity of a sauce is limited, and when too much oil is added, it "breaks" (separates back) into two layers.

Mayonnaise is an emulsion of oil droplets suspended in a base primarily composed of egg yolk, lemon juice or vinegar, and water. (Other ingredients are often added for flavor.) The oil and lemon juice or vinegar are stabilized by lecithin, a component of egg yolk. The lecithin molecule has an ionic head and a nonpolar tail, which cause micelles to form.

Adding soap to a "broken emulsion" (one that has separated into its insoluble components) stabilizes the emulsion again. However, unless the emulsifying agent is in a high concentration, it is not effective and so it is often necessary to add small amounts of the "broken emulsion" to a larger quantity of the emulsifying agent. This process allows micelles to be formed as necessary for the emulsion to reform. For example, mayonnaise that has separated must be added and mixed in small portions to beaten egg yolk. When the sequence of addition is reversed, the emulsifying agent is not in a high enough concentration to produce micelles effectively.

In margarine, water droplets are suspended in fat. Some margarine brands, advertised as "light" margarine, have more water than regular margarine. To compare the water content of different brands, the margarine can be melted. Upon melting, the emulsion of water in fat breaks down and the water and fat layers separate, allowing students to see that the amount of water varies in different brands.

Milk is an emulsion of milk fat suspended in water. Raw milk tends to separate into two layers with the fat-rich cream rising to the top. Homogenization of milk prevents the separation process through a high pressure mechanical process that breaks the fat particles into very small particles.

Some substances behave similarly to emulsifying agents, but their effect is more temporary. For example, solid paprika and dry mustard help stabilize French dressing by adhering to the surfaces of the oil droplets. Honey and sugar syrup are used to thicken some salad dressings; shaking suspends the oil droplets and their coalescing is slowed by the viscosity of the liquid.

Answers to Student Questions

Part A

Step 1

a. *The color of the vegetable oil*
b. *Density*
c. *Water sinks to the bottom because it is more dense than the oil.*

Step 2

a. *None of the added water dissolved in the oil because the height of the oil layer did not change.*
b. *Water is a polar covalent substance and oil is a nonpolar covalent substance. A general rule of thumb for the solubility of covalent substances in one another is that "like dissolves like." Since water is polar and oil is not, they are not alike; thus they aren't expected to dissolve in one another.*

Step 4

See table below.

Additive	Polarity	Layer Dissolved In
tincture of iodine	mixture of both polar and nonpolar species	water (red-brown) and oil (light red-purple)
food color	polar	water
mineral oil	nonpolar	oil
baby oil	nonpolar	oil
rubbing alcohol	polar	mostly water; some oil
soft drink	polar	water

Step 5

a. *The mixture takes much longer to separate into layers.*

b. *Soap or detergent molecules have a long nonpolar portion, which is attracted to the oil molecules, as well as an ionic portion, which is attracted to water molecules. The result is that some of the oil gets suspended in the water, forming what is known as an emulsion.*

Step 6

All include water, vinegar, some type of mixture of oils, and various flavors. The creamy type also includes either egg yolks or mayonnaise (as a source of egg yolk), which serves as an emulsifier.

Part B

Step 1

Answers will vary with the type of hand cream used, but typically water, glycerine, and a wax or oil are high on the list.

Step 2

a. *The mixture may appear curdled, and in some cases a thin layer of oil may be visible floating on the surface of the water. There is both a wet and an "oily" feeling on your finger.*

b. *The water-soluble components dissolved in water, leaving the oils undissolved (and floating on top since they are less dense than water).*

c. *Hand cream can hold only so much water before the water and oil components separate.*

Step 3

a. *The mixture is no longer curdled in appearance; it now appears smooth.*

b. *The emulsion was reformed by slowly stirring the separated mixture with the liquid detergent. The "repaired" emulsion of water, hand cream, and detergent is less viscous due to the increased amount of water present.*

Step 4

a. *The mixture again appears curdled; drops are dispersed in the oil; and there is both a wet and an "oily" feeling on your finger.*

b. *The oil-soluble components dissolved in the oil and the water was left in the more dense layer or in drops.*

c. Hand cream can hold only so much oil before the oil and water components separate.

Step 5

a. As the oily mixture is added to the soap, the curdled mixture becomes smooth and the layers combine.

b. The emulsion was reformed by slowly stirring the separated mixture into the liquid detergent. The "repaired" emulsion of oil, hand cream, and detergent is more viscous than the repaired emulsion of water, hand cream, and detergent.

References

Corriher, S.O. *Cookwise;* HarperCollins: New York, 1997.

Glover, A.D.; Kolb, K.E. Demonstrating What 'Light' Margarine Means. *J. Chem. Educ.,* **1991**, *68* (8), 654.

Curdled Milk

Overview

In this lesson, students explore the effects of an acid, an enzyme, and temperature on the curding behavior of milk.

Key Concepts

- catalysts
- chemical reactions
- denaturation
- enzymes
- experimental design
- proteins

National Science Education Standards

Science as Inquiry

Abilities Necessary to Do Scientific Inquiry

- *Students use thermometers to monitor the temperature of milk. (5–8, 9–12)*
- *Students develop descriptions and explanations about the effects of rennin on milk at different temperatures. (5–8)*
- *Students use logic and evidence to formulate explanations about the effect of adding rennin and vinegar to milk at different temperatures. (9–12)*
- *Students pose testable questions and design experiments to answer these questions. (5–8, 9–12)*

Physical Science

Properties and Changes of Properties in Matter

- *Students observe that rennin and vinegar chemically react with milk to form curds and whey, which have different characteristic properties. (5–8)*

Chemical Reactions

- *Students discover that rennin is an enzyme that catalyzes the coagulation of the casein in milk to form curds. This denaturation of the protein in milk requires temperatures between 10°C (50°F) and 45°C (110°F). (9–12)*

Life Science

Structure and Function in Living Systems

- *By investigating rennin (an enzyme obtained from the stomach lining of calves), students discover the role of enzymes in the digestive system. (5–8)*

The Cell

- *By investigating rennin, students discover the role of enzymes in breaking down and synthesizing food molecules during digestion. (9–12)*

Student Exploration

What happens when rennin (an enzyme) or vinegar (an acid) is added to milk? How does temperature affect what happens?

Materials

- 4 half-tablets of rennin (sold as Junket® Rennet tablets), crushed

➤ *Junket can usually be found near powdered gelatin in the grocery store.*

- 120 mL (½ cup) whole, 2%, or skim milk
- 1 or more saucepans or appropriate heat-safe containers
- stove, hot plate, or microwave
- thermometer capable of reading up to 80°C (180°F)
- tablespoon and teaspoon measures
- 4 clear plastic cups
- container of ice water large enough to hold 1 clear plastic cup
- paper towels
- 15 mL (1 tablespoon) vinegar
- 1 mL (about ¼ teaspoon) baking soda
- water
- stirring rod or spoon
- items to stick together in step 5, such as pieces of paper or craft sticks

Procedure

➤ *Use a clean heat-safe container in each step.*

1. In the heat-safe container, gently warm (do not boil) a 30-mL (1-ounce) sample of milk to approximately 45°C (110°F). Stir one half-tablet (crushed) of rennin into the milk. Let sit for 5–10 minutes and then record your observations. Stir the mixture. *How does the addition of rennin affect the milk?* Gather any solid material (curds) with a paper towel and squeeze out and discard any excess liquid (whey). Save any curds you collect for step 5.

2. Heat a second 30-mL (1-ounce) sample of milk to approximately 80°C (180°F). Stir one half-tablet of rennin (crushed) into the milk. Let the mixture sit for 5–10 minutes and then stir. *Compare the observed results to those in step 1.* Cool the milk to 45°C (110°F) and compare to results from the heated sample. Add another half-tablet of rennin (crushed); let sit for a few minutes and then stir. *Observe and compare your results. Make a claim about the activity of rennin based on the results of this step.* Again, gather and save any curds for step 5.

3. Put a third 30-mL (1-ounce) sample of milk into a clear plastic cup and place the cup in the container of ice water. When the milk cools below 10°C (50°F), add one half-tablet of rennin (crushed) and stir. Wait 5–10 minutes and record your observations. Pour the mixture into the heat-safe container

and gently warm it to 45°C (110°F). *Make a claim about the activity of rennin based on this new evidence.* Again, gather and save any curds for step 5.

4. Place a fourth 30-mL (1-ounce) sample of milk into a clear plastic cup and stir in 15 mL (1 tablespoon) vinegar. *Compare your observations with those from step 1.* Again, gather and save any curds for step 5.

5. Separate the curds into small pieces. Add 15 mL (1 tablespoon) water and 1 mL (about ¼ teaspoon) baking soda. Stir thoroughly. *What happens?* Sandwich some of the product between two craft sticks, index cards, or pieces of paper. Allow to dry and make observations. *Make a guess as to what this product is.* Do a web search to check whether you are correct.

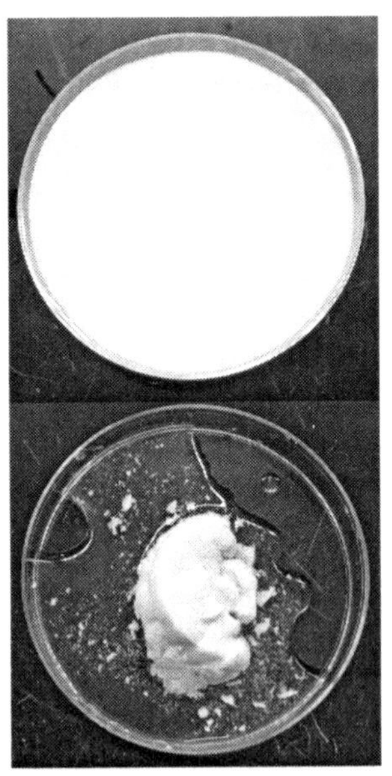

Fresh (top) and curdled (bottom) milk in petri dishes

Instructor Notes

Tips and Instructional Strategies

- You may want to divide the class into three groups and have each group use a different type of milk from among those listed in the Materials section. Challenge students to research the differences among the three types of milk and use what they learn to explain any differences in their results. (Typically, higher fat milk will result in stickier, less separated curds.)
- You may also want to have additional student groups try UHT milk and canned (evaporated) milk. UHT milk (or ultra high temperature processed milk) is stored and sold at room temperature and is often packaged in a multilayered box. Challenge students to determine whether UHT milk or canned milk could be used to make cheese. Since neither UHT milk nor canned milk reacts with rennin, students should conclude that these products could not be used to make cheese. For older students, lead them to the conclusion that heating is part of the production process for these two milk types and that heat has altered the milk protein.
- To save time you may want to heat a larger amount of milk to 50°C (120°F) before beginning the activity.
- Be sure students discover the temperature dependence of the rennin.
- Students should compare the curd size and texture produced with rennin to the smaller and drier curd produced in the reaction of vinegar and milk. The time for enzymatic curd formation with rennin is slower (5–10 minutes) than the reaction with vinegar.
- You may want to have students investigate the important roles that enzymes such as rennin and pepsin play in digestion, and, in particular, in the nutrition of young mammals.
- Challenge students to pose testable questions based on what they've done and learned in the lesson. They can design experiments and collect evidence to answer these questions, formulate claims about their findings, and, if time allows, present and defend their claims with their group or class. While we strongly recommend that students develop their own testable questions, you may need to seed the discussion with possible questions such as these: How do the results vary with the amount of rennin (or ratio of rennin to milk) used? How would the results vary if different dairy products (such as buttermilk, heavy whipping cream, half-and-half, or melted ice cream) were used in place of the milk?

Explanation

This activity explores some properties of milk and how milk is affected by the enzyme rennin and by vinegar. Milk contains the globular protein casein, as well as milk fat. Like all globular proteins, the three-dimensional structure of casein is altered when it is heated or when the pH level (acidity) is changed beyond a certain range. This change in structure is called denaturation. When proteins denature, they change their shape, which causes them to also lose their enzymatic activity.

The milk fat, which would normally separate as a layer of cream floating on the top of the milk, is thoroughly suspended in a process called homogenization to form a colloidal suspension. Thus raw milk is "homogenized" to retard separation.

When milk spoils, the lactose sugar in milk is oxidized to lactic acid (which is responsible for the sour taste) and the lactic acid causes denaturation of the casein protein. The denaturation of a globular protein decreases its solubility and the result is the formation of curds. (The remaining liquid is called whey, thus the "curds and whey" eaten by Little Miss Muffet.) In this activity, vinegar (a dilute solution of acetic acid) causes the milk curds to form.

Rennin, an enzyme, also causes the casein to coagulate, forming curds. Enzymes are biological catalysts, which means they change the speed of chemical reactions within living systems. Enzymes are globular proteins and their enzymatic activity is very much dependent upon their three-dimensional structure. Enzyme structure can be altered (denatured) by temperature, so at both higher and lower temperatures, enzymes can become inactive. Typically, the active temperature range of enzymes in living bodies is around body temperature, (for humans, 37°C or 98.6°F). In this activity, the rennin causes coagulation of casein at temperatures between 10°C (50°F) and 45°C (110°F), but the rennin becomes inactivated at lower and higher temperatures.

The cheese-making process typically involves the reaction of rennin and milk, in which the formation and coagulation of the curd and separation of the curd from the whey occur. Rennin, a protease enzyme, cleaves the peptide bond of the milk protein casein between phenylalanine and methionine. This cleavage separates the hydrophobic and hydrophilic part of the protein. The hydrophobic groups coagulate, causing curd formation and precipitation. The larger curds formed in rennin precipitation contain casein, whey protein, fat, lactose, and minerals. In addition to its nutritional applications, casein is used industrially in paints, white glues, plastics, fibers, and sizing to treat paper.

Answers to Student Questions

Step 1

Large, fluffy, and sour-smelling curds form. At approximately 45°C (110°F), rennin causes coagulation of the protein, casein.

Step 2

a. Curds do not form when rennin is added to the mixture at 80°C (180°F).

b. Curds do not form upon cooling the mixture down to 45°C (110°F, the same temperature as in step 1).

c. Curds do form at 45°C (110°F) when more rennin is added.

d. Rennin is rendered inactive at high temperatures. When the milk is heated to the higher temperature, the rennin that was added permanently loses its ability to cause coagulation of casein. The increased temperature denatures the protein and so deactivates its enzymatic activity. Additional rennin (which has not been heated) must be added to cause coagulation.

Step 3

There is no apparent formation of curds when the temperature is below 10°C (50°F). A temperature somewhere between 10°C (50°F) and 45°C (110°F) is necessary for rennin to cause coagulation of casein to form curds.

Step 4

Curds form faster, but are smaller and less fluffy than in step 1. Making the solution acidic causes the casein to denature and results in its coagulation.

Step 5

a. The mixture turns into a sticky material.

b. The items become "glued" together. The product that has been produced is a casein glue.

Reference

Sarquis, A.M., Sarquis, J.L., Eds. Curdling of Milk. *Fun with Chemistry: A Guidebook of K–12 Activities,* Vol. 1; Institute for Chemical Education: Madison, WI, **1993;** pp 63–66.

Investigating Reusable Heat Packs

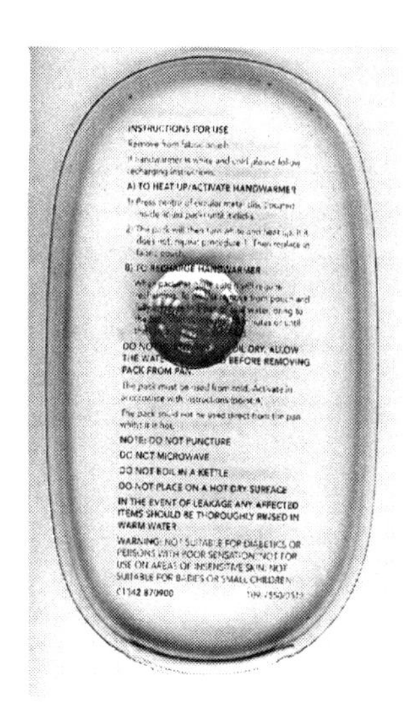

Overview

Can a small bag containing a solution act as a source of heat at a moment's notice? Students discover how a reusable heat pack provides heat and measure the amount of heat given off. In the culminating teacher demonstration, the class makes a heat pack.

Key Concepts

- calorimetry
- changes of state
- concentration
- conservation of energy
- crystallization
- energy changes
- exothermic processes
- experimental design
- heat
- heat energy
- insulators
- saturated solutions
- solutes
- solutions
- solvents
- specific heat
- supersaturated solutions
- temperature

National Science Education Standards

Science as Inquiry

Abilities Necessary to Do Scientific Inquiry

- *Students use a calorimeter to determine the amount of heat produced by the heat pack. (5–8, 9–12)*
- *Students use mathematical equations to determine the amount of heat produced in the crystallization process. (5–8, 9–12)*
- *Students see how mathematical tools and models guide and improve questioning, gathering data, constructing explanations, and communicating results. (5–8, 9–12)*
- *Students determine what data they will need to collect in order to calculate the amount of heat given off by a heat pack. (5–8, 9–12)*
- *Students design and conduct an experiment, using a control and testing one or more variables, to determine how to slow down the loss of heat from a heat pack. (5–8, 9–12)*
- *Students pose testable questions and design experiments to answer these questions. (5–8, 9–12)*

Physical Science

Structure and Properties of Matter

- *Students investigate the crystallization process and explore the use of seed crystals to initiate the process. (9–12)*

Conservation of Energy and the Increase in Disorder

- *The lesson illustrates a consequence of the second law of thermodynamics. Crystallization increases the order (reduces the entropy) of the contents of the heat pack, so to maintain an overall increase of entropy, energy is released to the surroundings as heat. (9–12)*
- *Students learn that heat is a form of energy produced by the motion of small particles of matter and that temperature is a measure of a system's ability to gain or lose heat. (9–12)*

Interactions of Energy and Matter

- *Students explore the effectiveness of various materials as insulators and discover that insulation slows the transfer of energy from the heat pack to the environment. (9–12)*

Transfer of Energy

- *Students learn that heat moves in predictable ways, in this case from the heat pack to its surroundings. (5–8)*

Part A: Teacher Demonstration

How does a heat pack work?

Materials

- reusable heat pack
- overhead projector

Procedure

➤ *Make sure the heat pack ingredients are in solution form, not crystalline form.*

1. Ask the students to name several different ways to generate heat. Then find out how many of these are "portable" sources. Show them the heat pack, asking if they know what it does.
2. Pass the heat pack around and ask students to make observations, cautioning them to handle the solution gently and not to touch the pack near the metal disk.
3. Place the bag on an overhead projector as you instruct the class to observe. Activate the heat pack by flexing the metal disk gently. After the students observe the crystallization process on the overhead, pass the bag from student to student so that all can observe.
4. Ask questions such as, "Where does the heat comes from? How can the bag be reused?" Demonstrate the redissolving process if time allows.

Part B: Student Exploration

How much heat does the heat pack give off? How can we keep the pack from losing its heat?

Materials

- reusable heat pack
- water
- thermometer
- graduated cylinder
- 480-mL (16-ounce) or larger Styrofoam® cup
- water
- other materials as needed for this student-designed experiment

Procedure

1. Decide what measurements you need to make in order to calculate the amount of heat (in Joules) given off by a heat pack.

➤ *Use the equation $q = m \times \Delta T \times C_p$ where q is the heat released in Joules, m is the mass of the water in grams, ΔT is the change in temperature in Kelvin, and C_p is the specific heat capacity. (For water, C_p is 4.18 J/g•K.) To convert °C to K, use the formula °C + 273 = K.*

2. Devise a plan to collect this data. Use a large Styrofoam cup as a calorimeter. Implement your plan, collect the necessary data, and calculate the heat released by the activated heat pack. *How much heat would be needed to return the contents of the heat pack back to a solution?*

3. *Make a list of the factors that affect the time required for the fully heated pack to cool to room temperature.* Design a method to delay this heat loss for as long as possible. Test your method and be sure your experiment includes a control.

Part C: Teacher Demonstration

Can we make it ourselves? Manufacture your own heat pack using a supersaturated solution of sodium acetate.

Materials

- sodium acetate trihydrate ($NaC_2H_3O_2 \cdot 3H_2O$)
- distilled water
- 1-L flask
- hot-water bath, made from the following:
  - large container or pot
  - hot plate or other heat source
- glass pie pan, crystallizing dish, or piece of clean hardboard measuring about 30 cm × 30 cm (about 12 inches × 12 inches)
- 2 clean 250-mL flasks or zipper-type plastic bags
- petri dishes
- watch glass or foil
- overhead projector
- table salt (NaCl) and other crystalline solids

Preparation

1. Combine 350 g (about 1½ cups) sodium acetate trihydrate ($NaC_2H_3O_2 \cdot 3H_2O$) and 100 mL distilled water in the flask.
2. Place the flask with its contents in the hot-water bath, cover with a watch glass or foil to minimize water loss, and heat the water to boiling. The flask can be swirled occasionally to speed the dissolution, but caution must be used, as the flask and the steam will be very hot. Once all of the solid has dissolved, turn off the heat source and allow the setup to cool to room temperature before removing the flask from the hot-water bath. Keep the mouth of the flask covered with a watch glass or foil until ready to use.

Procedure

1. Pour about ⅓ of the room-temperature supersaturated sodium acetate solution into either a clean flask or a zipper-type plastic bag container. Have volunteers feel the sides of the container and report what they feel to the class. Ask the students, "What is required to initiate crystallization?"
2. Add 1 or 2 small "seed" crystals of the sodium acetate trihydrate to the flask or bag and observe. Have the volunteers feel the container again and report their observations. Turn the container sideways to see if any liquid remains.

3. To show a dramatic example of crystallization, very slowly pour about half of the remaining supersaturated sodium acetate solution in the flask onto a few crystals of sodium acetate that are lying on a glass pie pan, crystallizing dish, or piece of clean hardboard or tabletop.

➤ *Crystallization of the supersaturated solution occurs as soon as the solution comes in contact with the seed crystal, forming a mound or column of white solid sodium acetate. The lip of the flask or container must not come into contact with the crystallizing sodium acetate, since this would cause the rest of the solution to crystallize while still in the flask or container. Uncontaminated solutions of sodium acetate can be reused many times. They should be stored covered, as the supersaturated solution, instead of in the crystalline form.*

Restore the crystals to a supersaturated solution by covering the mouth of the flask with a watch glass or foil and reheating the flask in a hot-water bath. (If the solid is in a plastic bag, transfer it to a flask first.) Be sure that no undissolved crystals remain. Turn off the heat source and allow the flask and its contents to cool to room temperature before removing from the hot water bath. Leave the flask covered until use. Small amounts of water may have to be added to compensate for loss of water due to evaporation when heating.

Instructor Notes

Tips and Instructional Strategies

- Discuss the terminology of solutions; for example, solute, solvent, unsaturated, saturated, and supersaturated.
- According to the manufacturer, the heat pack should always be stored in the solution form, not the crystalline form. Place the crystallized heat packs in boiling water for about 10 minutes to return to solution. Allow 45 minutes for the heat packs to cool to room temperature.
- For Part B, you may want to have students in lower grades calculate the heat released in calories, rather than Joules, according to the definition: 1 calorie equals the energy needed to raise the temperature of 1 gram of water by 1°C. (Specific heat for water in calories is 1.00 cal/g•°C.) Students in lower grades may also need hints on how to devise a plan to collect the necessary data.
- For Part B, step 2, students are likely to find that the maximum temperature is reached in 30–40 minutes when using 270 mL water. Note that stirring the water in the calorimeter will make a difference in the data collected. The maximum temperature will be reached sooner if the water is stirred.
- For Part B, step 3, discuss experimental design as a class. Emphasize the importance of controlling variables. Establish what the variables would be for each experiment proposed, and discuss how students would control these variables. You may want to develop a procedure as a class, assign different ways of insulating to different groups, and have one untreated heat pack serve as a control for the class. Alternatively, each group could develop and carry out their own experimental procedure.
- For Part C, prepare extra flasks of supersaturated sodium acetate solution, just in case. Flask contents may crystallize if jostled.
- In Part C, step 2, a small amount of solution may remain above the crystals. If observed, point out that a common misconception is that all of the solute crystallizes out of solution. Actually, only the "extra" solute crystallizes. The remaining solution is a saturated solution.
- You can also show the students that you can form intricate sculptures using a wash bottle of supersaturated sodium acetate to deliver the solution onto the seed crystals. (If the tip of the wash bottle comes into contact with the crystals, the entire contents of the bottle will crystallize.)
- Challenge students to pose testable questions based on what they've done and learned in the lesson. They can design experiments and collect evidence to answer these questions, formulate claims about their findings, and, if time allows, present and defend their claims with their group or class. While we strongly recommend that students develop their own testable questions, you may need to seed the discussion with possible questions such as these: Other than bending the metal disk as in Part A, can something be done to

the heat pack to cause the excess sodium acetate to precipitate from the supersaturated solution without cutting the bag open or otherwise destroying it? What effect does the starting temperature of the water have on the amount of heat released by the heat pack (as determined in Part B)? Can crystals of another solid be used to seed the crystallization process in Part C?

Explanation

The heat pack contains a supersaturated solution of the salt sodium acetate in water. A supersaturated solution is one in which there is more solute (sodium acetate) dissolved in the solvent (water) than would normally be possible at a given temperature. This is accomplished by heating the solution to a higher temperature and allowing it to slowly cool. With most salts, the extra solute that dissolved at the higher temperature crystallizes out of solution as it cools. However, with substances that form a supersaturated solution, this extra solute remains dissolved. Honey is an example of a supersaturated solution of sugar in water.

A supersaturated solution is inherently unstable but remains as a solution until something initiates crystallization. You may have seen jars of honey that became crystallized. In the heat pack, the flexing of the metal disk creates a shock wave that is sufficient to initiate crystallization. Once this occurs, the supersaturated solution immediately crystallizes to form the more stable solid. Heat is given off as the solution crystallizes. Supersaturated solutions can also be made to crystallize by adding a "seed" crystal, a crystal of the same solid that is dissolved in the supersaturated solution. Typically, crystals of different substances will not initiate crystallization.

Temperature is related to the average KE of the molecules and is measured in degrees C, K, F. Heat is a measure of total energy in a substance (PE and KE) measured in J or cal. Temperature is a measure of how hot or cold a substance is relative to another substance. Heat energy is transferred between two systems as a result of temperature difference, flowing in the direction of lower temperature.

The crystallization process is reversible. If the crystallized sample is heated, it goes back into solution. The heat absorbed by the crystallized heat pack to get the salt back into solution is released when the sodium acetate recrystallizes. (When sugar crystallizes in honey, it also can be made to go back into solution by heating.)

The amount of heat, q, transferred by the hot pack can be calculated from:

$$q = m \times \Delta T \times C_p$$

where m is the mass of the water used in grams, ΔT is the change in temperature in Celsius degrees or Kelvin*, and C_p is the specific heat capacity. (For water, C_p is 4.18 J/g•K or 1.00 cal/g•°C.) *Note: ΔT in °C is equal to the ΔT in K.

Answers to Student Questions

Part B

Step 2

a. *Calculate the ΔT using the starting and ending temperatures. Calculate the heat released in Joules (q) in the crystallization process using the equation below where m is the mass of the water used in grams and C_p is the specific heat capacity. (For water, C_p is 4.18 J/g•K.)*

$$q = m \times \Delta T \times C_p$$

b. *Since energy is conserved, the same amount of heat that was released when sodium acetate crystallized must be absorbed to return the solid to solution.*

Step 3

Factors include the size of the temperature change observed in the heat pack, the room temperature, and the type and amount of insulation.

Reference

Sarquis A.M., Sarquis, J.L., Eds. Crystallization of a Supersaturated Solution. *Fun with Chemistry: A Guidebook of K–12 Activities,* Vol. 2; Institute for Chemical Education: Madison, WI, 1993; pp 287–291.

Dissolving Energy

Overview

In this activity, students investigate temperature changes that occur when various household substances dissolve in water.

Key Concepts

- anhydrous substances
- dissolving
- experimental design
- heat of solution
- solutions
- water of hydration

National Science Education Standards

Science as Inquiry

Abilities Necessary to Do Scientific Inquiry

- *Students use thermometers to determine the temperature changes that occur when common household substances are dissolved in water. (5–8, 9–12)*
- *Students use logic and evidence to formulate explanations about how dissolving various amounts of solid in the same amount of water affects the temperature of the solutions. (5–8, 9–12)*
- *Students gain experience in identifying and controlling variables as they plan and conduct experiments to investigate the heat of solution of various solids in water. (9–12)*
- *In Part A, students design and conduct an experiment to determine whether the starting temperature of the water affects the temperature change observed. (5–8, 9–12)*
- *In Part B, students design and conduct an experiment to determine whether or not the ratio of solid to water affects the temperature change. (9–12)*
- *In Part C, students use logic and evidence to formulate explanations about the differences in the temperature changes resulting from dissolving Lite Salt and table salt. (5–8, 9–12)*

Physical Science

Properties and Changes of Properties in Matter

- *In Part C, students observe that dissolving Lite Salt in water produces a different temperature change than dissolving table salt and learn that the different substances in these products account for that difference. (5–8)*

Structure and Properties of Matter

- *In Part C, by dissolving Lite Salt and table salt in water, students observe that these substances have different heats of solution. They learn that Lite Salt contains not only sodium chloride, but also potassium chloride, which has a heat of solution that is more endothermic than that of sodium chloride. (9–12)*

Transfer of Energy

- *Students discover that when a solute dissolves in water, a noticeable change in temperature can occur and they can feel this heat transfer as well as measure the temperature change with a thermometer. (5–8)*

Chemical Reactions

- *Students observe and conclude that some chemical reactions and other processes that involve the breaking of bonds and the formation of hydrated ions during the dissolving process can release or consume energy. (9–12)*

Part A: Student Exploration

Do temperature changes occur when common household substances dissolve in water?

Materials

- 100-mL graduated cylinder or measuring cup
- tablespoon measure
- 2–3 small plastic cups
- stirrer
- water
- baking soda (sodium bicarbonate, $NaHCO_3$)
- powdered laundry detergent (for example, Tide® or Cheer®)
- table salt (sodium chloride, NaCl)
- washing soda (sodium carbonate, Na_2CO_3)

Washing soda is a severe eye irritant and minor skin irritant. Goggles must be worn at all times, and gloves are recommended. Washing soda may be harmful if swallowed or inhaled.

- thermometer with a scale that includes room temperature and allows measuring temperature increments of 2°C
- other materials as needed for the student-designed experiment in step 3

Procedure

Because some household solids are especially hazardous, confine your exploration to those listed in the Materials section.

1. Pour 60 mL (¼ cup) room-temperature water into a cup and record the temperature. Add 60 mL (4 level tablespoons) baking soda to the water all at once. Stir and record the temperature. (Do not stir with the thermometer.) Continue stirring and recording the temperature until a maximum or minimum temperature is observed. When no further change in temperature is observed, discontinue stirring and recording after 2–3 minutes. *What evidence of change do you observe in the dissolving process?*

2. Repeat step 1 using powdered laundry detergent in place of baking soda. *How does dissolving the detergent differ from dissolving baking soda?* Repeat with table salt and then with washing soda. *Formulate a claim about one or more of the systems you have studied and substantiate your claim with evidence you have collected.*

3. Design an experiment to determine whether the starting temperature of the water affects the temperature change observed.

Part B: Student Exploration

Use Epsom salt and anhydrous magnesium sulfate to illustrate the heating and cooling effects that can result when a salt is dissolved in water.

Materials

- 60 mL (¼ cup) Epsom salt (magnesium sulfate heptahydrate, $MgSO_4 \cdot 7H_2O$)
- 60 mL (¼ cup) anhydrous magnesium sulfate ($MgSO_4$)
- 2 small zipper-type plastic bags
- water
- 100-mL graduated cylinder
- permanent marker
- other materials as needed for the student-designed experiment in step 3

Procedure

1. Label one plastic bag Epsom salt ($MgSO_4 \cdot 7H_2O$) and the other anhydrous magnesium sulfate ($MgSO_4$). Place 60 mL (¼ cup) of the appropriate solid into each labeled bag.
2. Measure out 60 mL (¼ cup) room-temperature water. Record its temperature. Pour the water into one of the solid-containing bags. Seal the bag and shake to mix. Record the highest or lowest temperature of the solution after mixing. Repeat with the other solid and compare results. *Make a claim about differences in your observations of the dissolution of the anhydrous and hydrated forms of magnesium sulfate and substantiate the claim with the data you have collected. What chemical species do you think are present in solution when anhydrous magnesium sulfate dissolves? What chemical species do you think are present in solution when Epsom salt dissolves?*
3. Design an experiment to determine whether or not the ratio of solid to water affects the change in temperature.

Part C: Student Exploration

Explore the differences between two flavor-enhancing products, table salt and Lite Salt™, by dissolving each in water.

Materials

- tablespoon, measuring cup, or graduated cylinder
- 30 mL (⅛ cup or 2 level tablespoons) table salt
- 30 mL (⅛ cup or 2 level tablespoons) Lite Salt
- water
- 2 small cups
- stirrer
- thermometer

Procedure

1. Place 60 mL (¼ cup) room-temperature water into a cup and record the temperature. Add 30 mL (⅛ cup) table salt to the water all at once and stir. (Do not stir with the thermometer.) *Record the highest or lowest temperature reached.*
2. Repeat step 1 using the Lite Salt in place of the table salt. *Read the label for the Lite Salt and record its composition. Use this information to interpret your observations. Why does the Lite Salt behave differently than table salt?*

Instructor Notes

Tips and Instructional Strategies

- This activity can be used when studying solutions and endothermic and exothermic processes. You may want to do Parts B and C with older students only.
- To prepare dehydrated Epsom salt for Part B, heat 120 mL (½ cup) Epsom salt (magnesium sulfate heptahydrate, $MgSO_4{\cdot}7H_2O$) in an aluminum pie pan for 30 minutes at 450°F (230°C). Remove from the oven and allow to cool for 10 minutes. Cover the dehydrated Epsom salt with plastic wrap and use a hard flat object to crush the salt into small pieces. Store the salt in a sealed container to minimize rehydration. This dehydration step can be done several days in advance of the activity or you may want to have older students do the dehydration as part of the activity. By doing this step themselves, students could determine the percentage of water in Epsom salt experimentally and compare their results with the actual percentage.
- Part C could also be done using plastic bags, as in Part B. The bag in which the Lite Salt dissolves will feel noticeably cooler than the one in which the table salt dissolves.
- As an extension to Part C, step 2, you may want to challenge students to predict what would happen if pure KCl were dissolved instead of the Lite Salt.
- You may want to challenge the students to calculate the amount of heat energy that is released or absorbed during solution formation. Have students record the volume of water used to the nearest milliliter. Knowing that the specific heat capacity of water is 1.00 cal/g•°C and using the water volume and the temperature change, students can calculate the amount of heat released or absorbed.
- For older students, you may want to demonstrate the dissolution of lithium chloride (LiCl) in water. Dissolving 42 g LiCl in 50 mL distilled water raises the temperature of the resulting solution to above 65°C (149°F). Follow appropriate safety precautions, as the hot solution can cause severe burns. You could also challenge students to find a periodic pattern for Li, Na, and K and ask them to predict what would happen with cesium chloride (CsCl).

Explanation

Many common household materials produce noticeable energy changes when dissolved in water. In this activity, students observe the energy released or absorbed by a solid dissolving in liquid water to form a solution.

In general, a solution consists of one substance, the solute, dissolved in another, the solvent. The solute is generally the component present in the smallest amount. In this activity, a variety of household solids (the solutes) were dissolved in water (the solvent). The temperature changes observed result from a transfer of energy that occurs during the dissolving process, either into (endothermic) or out from (exothermic) the resulting solution.

When baking soda ($NaHCO_3$) dissolves in water, the temperature noticeably decreases. For $NaHCO_3$ to dissolve in water, energy (in the form of heat) is transferred from the surroundings, such as your fingers, the thermometer, the container, or the air, into the $NaHCO_3(aq)$ solution. Since heat is leaving the fingers, a cooling sensation is perceived. In this endothermic process, the resulting solution has gained energy, and the energy change is represented by a positive number. When washing soda (Na_2CO_3) dissolves in water, heat is transferred from the solution to the surroundings. This is an exothermic process and the energy change is represented by a negative number.

Part B calls for the use of Epsom salt ($MgSO_4{\cdot}7H_2O$) and anhydrous $MgSO_4$. In Epsom salt, the $MgSO_4$ has water molecules (called water of hydration) in a definite ratio of water to salt—every magnesium sulfate formula unit has seven water molecules arranged about it, yielding crystals that have a characteristic shape different from that of anhydrous $MgSO_4$. The anhydrous $MgSO_4$ is prepared by heating the $MgSO_4{\cdot}7H_2O$, which drives off the water of hydration as water vapor. At 150°C, six of the water molecules are released. At 200°C, the seventh water molecule is released, leaving the anhydrous $MgSO_4$. In this activity, the Epsom salt is heated at about 230°C for 30 minutes, which removes all of the water of hydration.

The activity compares the heat changes that occur when $MgSO_4{\cdot}7H_2O$ and anhydrous $MgSO_4$ dissolve in water. While the same aqueous ions exist in the resulting solutions, there is a significant difference in the temperatures of the resulting solutions. These two situations can be described in the form of equations that include heat as a product in the first case and heat as a reactant in the second case.

Dissolving anhydrous $MgSO_4$ in water:

$$MgSO_4(s) \xrightarrow{H_2O} Mg^{2+}(aq) + SO_4^{2-}(aq) + heat$$

Dissolving Epsom salt in water:

$$MgSO_4{\cdot}7H_2O(s) + heat \xrightarrow{H_2O} Mg^{2+}(aq) + SO_4^{2-}(aq) + 7H_2O(l)$$

The following discussion uses the term hydration in two contexts. First, when a solid ionic compound dissolves in water, the ions are separated from one another. This is because the polar water molecules shield the positive charge on the cation from the negative charge on the anion, allowing the ions to dissolve. These ions are described as hydrated, meaning water molecules surround each oppositely charged ion, which keeps the ions from re-crystallizing as the solid salt. The symbols for hydrated ions are often followed by the (aq) designation.

In Epsom salt crystals, "water of hydration" exists within the solid crystal structure; these water molecules are a definite part of the crystal, seven per each magnesium and sulfate ion. To understand the cause of the temperature differences observed when these two salts are dissolved, we have to be careful not to mix up "hydrated ions in solution" with the "water of hydration" in the crystal.

When anhydrous $MgSO_4$ dissolves in water, the resulting ions, Mg^{2+}(aq) and SO_4^{2-}(aq), are surrounded by water molecules, which allow these ions to stay separated in solution; this dissolving process releases energy in the form of heat. (The heat of solution, ΔH, is –84.9 kJ/mol.) When $MgSO_4{\cdot}7H_2O$ dissolves, once again the Mg^{2+}(aq) and SO_4^{2-}(aq) are surrounded by water molecules and energy is released. But another factor is involved—energy is required in order to overcome the attractive forces between the water of hydration and the Mg^{2+} and SO_4^{2-} ions in the Epsom salt crystal. The evidence for the overall endothermic nature of the dissolution of Epsom salt (ΔH = +16.2kJ/mol) is the cooling of the resulting solution. This is because more energy must be taken in from the surroundings to accomplish the loss of the water of hydration from the crystalline structure than the energy that is released by the formation of hydrated ions in solution.

In Part C, table salt is mostly sodium chloride (NaCl). The container label of the Lite Salt lists NaCl as the first ingredient and potassium chloride (KCl) as the second. The heat of solution of KCl is more endothermic (+17.24 kJ/mole) than the heat of solution of NaCl (+3.9kJ/mole). Thus, a slightly larger cooling effect is expected for the Lite Salt.

If you demonstrated the dissolution of lithium chloride (LiCl), you noted that the heat of solution is highly exothermic (ΔH = –37.1kJ/mole). The difference between the heat of solution of NaCl, KCl, and LiCl is due to the smaller size of Li^+ compared to Na^+ or K^+.

Answers to Student Questions

Part A

Step 1

Evidence of change is the temperature decreases (energy is absorbed) and the amount of solid baking soda decreases.

Step 2

a. *At least some of each solid dissolves in both cases; with the detergent, a temperature increase occurs (energy is released).*
b. *As different solids dissolve in water, different amounts of heat may be absorbed or released. Whether heat is released or absorbed depends on the identity of the solid. For example, dissolving baking soda absorbs more heat than dissolving table salt. Dissolving washing soda releases heat.*

Part B

Step 2

a. *A temperature increase occurs when the anhydrous $MgSO_4$ dissolves, and a temperature decrease occurs when Epsom salt dissolves. Due to the presence of water of hydration in the Epsom salt, it has a different heat of solution than anhydrous $MgSO_4$.*

b. *When anhydrous $MgSO_4$ dissolves, Mg^{2+}(aq) and SO_4^{2-}(aq) are present. (Note that these are hydrated ions.)*

c. *The same species are present as when anhydrous $MgSO_4$ dissolves.*

Part C

Step 1

For table salt, a decrease of about 2°C is typically observed.

Step 2

For Lite Salt, a decrease of about 8°C is typically observed. The Lite Salt behaves differently from the table salt, which is sodium chloride (NaCl), because in addition to NaCl, the Lite Salt also contains potassium chloride (KCl). In addition, students can infer that the K^+ is responsible for the greater degree of cooling.

References

Rybolt, T.R.; Mebane, R.C. Can Salt Be Used for Heating and Cooling? *Environmental Experiments about Renewable Energy*; Hillside, NJ: Enslow Publishers, Inc.,1994.

Sarquis, A.M., Sarquis, J.L., Eds. Energy Changes of Everyday Materials. *Fun with Chemistry: A Guidebook of K–12 Activities,* Vol. 2; Institute for Chemical Education: Madison, WI, 1993; pp 223–228.

Shakhashiri, B.Z. *Chemical Demonstrations: A Handbook for Teachers of Chemistry,* Vol. 1; University of Wisconsin: Madison, WI, 1983; pp 21–22.

Pencil Electrolysis

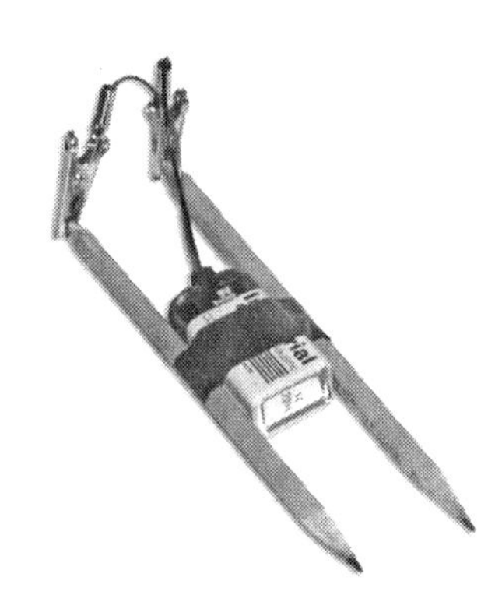

Overview

Water is considered to be a rather stable compound. Adding energy in the form of heat usually causes only a change in its state; the steam that results is still H_2O. Nevertheless, ordinary materials can be used to decompose water into its elements.

Key Concepts

- chemical changes
- compounds
- conductors
- decomposition reaction
- electricity
- electrolysis
- electrolytes
- elements
- water and its properties

National Science Education Standards

Science as Inquiry Standards

Abilities Necessary to Do Scientific Inquiry

- *Students observe the formation of gas bubbles at the positive and negative electrodes as water decomposes. (5–8, 9–12)*
- *From the production of gas bubbles, students infer that water is being decomposed. (5–8, 9–12)*

Physical Science

Transfer of Energy

- *Students observe a chemical change that involves the use of electrical energy to decompose water. (5–8)*

Structure and Properties of Matter

- *Students learn that water is made of molecules, which in turn are made up of hydrogen and oxygen atoms. (9–12)*

Chemical Reactions

- *The decomposition of water involves the transfer of electrons. (9–12)*

Interactions of Energy and Matter

- *Students find that it is necessary to add compounds such as Epsom salt or Glauber's salt to the distilled water to enable the conduction of electricity. (9–12)*

Teacher Preparation

Set up the apparatus and prepare the electrolyte solution.

Materials

- pencil electrolysis apparatus (includes 9-volt battery, 2 alligator clips, 9-volt battery T-cap, and 2 number 2 pencils)

Test pencils before use. Some inexpensive pencils do not contain enough graphite to allow them to serve as electrodes.

- water
- tape
- squirt bottle
- 1 of the following salts for an electrolyte solution:
  - magnesium sulfate heptahydrate, $MgSO_4 \cdot 7H_2O$ (Epsom salt)
  - sodium sulfate decahydrate, $Na_2SO_4 \cdot 10H_2O$ (Glauber's salt)

Do NOT substitute NaCl (table salt) for the listed salts. Using NaCl would result in the generation of highly toxic Cl_2 gas.

Procedure

1. Sharpen both ends of the pencils and assemble the electrolysis apparatus as shown in Figure 1.

After use, disconnect the alligator clips to prevent overheating and draining the battery. Make sure the alligator clips do not touch each other. (Attach them to the wooden part of the pencil during storage.)

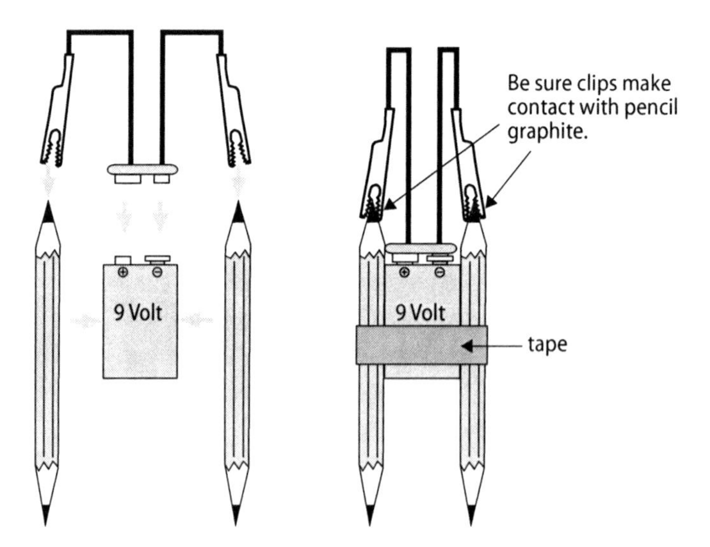

Figure 1: Assemble the pencil electrolysis apparatus.

2. Prepare about 50 mL of a saturated solution of one of the following salts to be used as an electrolyte solution in this activity.

- $MgSO_4 \cdot 7H_2O$ (Epsom salt)
- $Na_2SO_4 \cdot 10H_2O$ (Glauber's salt)

To do this, stir $MgSO_4 \cdot 7H_2O(s)$ or $Na_2SO_4 \cdot 10H_2O(s)$ into about 50 mL water until no more dissolves. Allow the mix to settle, and decant the saturated solution into a labeled squirt bottle.

Student Exploration

What does this contraption do?

epsom salt w/ water will work

Materials

- electrolysis apparatus
- shallow bowl or petri dish
- distilled water
- electrolyte solution in squirt bottle
- 5–10 drops of one of the following indicator solutions:
  - universal indicator
  - bromocresol green
  - red cabbage juice

Procedure

1. Pour distilled water to a depth of 0.5 to 1 cm in a clean, shallow bowl or petri dish.
2. Attach the battery T-cap to the battery and the alligator clips to the pencil tips.
3. Place the pencil tips (electrodes) of the electrolysis apparatus in the water. Look very closely at the submerged pencil tips. *Describe and explain your observations.* Remove the pencils.
4. Add about 5–10 drops of the indicator solution to the water so that you have an intensely colored solution, and again submerge the pencil tips. *Describe your observations.* Remove the pencils.
5. Squirt in some of the electrolysis solution and swirl it around. Place the pencil tips in the water. Observe carefully. *Record the color of the solution around each electrode. Describe what you see at each pencil tip. Is the same amount of gas being produced at both electrodes? On which side is hydrogen gas being produced? Why do you think so?*

Instructor Notes

Tips and Instructional Strategies

- Don't tell students what will happen. At the end of the activity, as students attempt to explain their observations, you may have to lead them to understand that they're "taking apart" water. Once students understand this idea, they may be able to identify which element is produced at which electrode, because they can reason that the side with twice as many bubbles must be the hydrogen side.
- Pencil electrolysis is also effective as a teacher demonstration if done on an overhead projector.
- If universal or bromocresol green indicator solution is not available, you can substitute red cabbage juice. This can be prepared by one of these methods:
    - Put approximately ¼ head of red cabbage into a blender or food processor. Add a little water and blend the cabbage into a slurry (1–3 minutes). Pour the slurry through a strainer, collecting the juice in a beaker.
    - Chop ¼ head of red cabbage and put it into a beaker. Add rubbing alcohol to cover the cabbage. Stir periodically. After about an hour, decant the liquid and discard the cabbage pieces.
- Make sure students understand the significance of the colors they see in step 5. The color changes because the pH around each electrode changes as the oxygen and hydrogen gases are produced. The pH around the cathode (negative electrode) becomes more basic because OH^- ions are produced along with $H_2(g)$. The pH around the anode (positive electrode) becomes more acidic because H^+ ions are produced along with $O_2(g)$.
- As a way for students to construct their own knowledge of the decomposition process, they can determine whether the electrode with the greater formation of a gas turns the indicator color due to excess OH^- or H^+ ions. This will allow them to identify that the electrode with greater formation of a gas is also the electrode where OH^- is produced, and the electrode where reduction occurs. Opposite arguments can be constructed for the reaction and behavior of the electrode where oxygen is formed.
- After the activity, be sure to disconnect the electrodes from the battery and to store alligator clips separately from the rest of the setup to avoid shorting out the batteries.

Explanation

Water is a chemical compound made from the elements hydrogen and oxygen in a 2:1 ratio. In this activity, water (a colorless, odorless liquid at standard temperature and pressure) is decomposed (broken apart) into its component elements, hydrogen and oxygen. Hydrogen (H_2) and oxygen (O_2) are both colorless, odorless gases at standard temperature and pressure.

The observed decomposition resulted from electrolysis (the use of electrical energy to decompose a compound). In step 3, students observed that distilled water is not a good conductor of electricity. It is necessary to add an electrolyte, such as Epsom salt (magnesium sulfate heptahydrate, $MgSO_4{\cdot}7H_2O$) or Glauber's salt (sodium sulfate decahydrate, $Na_2SO_4{\cdot}10H_2O$), to allow the electrical circuit to be completed through the solution so that the decomposition of the water in the solution can occur. The graphite in the pencils acts as the electrodes in this system. Placing the tips of these electrodes (pencils) into this electrolytic solution completes the circuit and evolution of H_2 and O_2 gas is observed at both electrodes.

The following two half-reactions represent the change taking place at the electrodes. The reduction half-reaction occurs at the cathode. The oxidation half-reaction occurs at the anode.

$$
\begin{array}{rcll}
4H_2O(l) + 4e^- & \rightarrow & 2H_2(g) + 4OH^-(aq) & \text{cathode} \\
2H_2O(l) & \rightarrow & O_2(g) + 4H^+(aq) + 4e^- & \text{anode} \\
\hline
2H_2O(l) & \rightarrow & 2H_2(g) + O_2(g) & \text{overall equation}
\end{array}
$$

Note that the equation stoichiometry predicts two moles of H_2 and one mole of O_2 will be evolved. Secondary reactions involving these reactive gases at the electrode surfaces may cause smaller amounts of the gases to actually be produced. For example, the black precipitate formed at the electrodes partially indicates these secondary reactions.

Adding an indicator solution provides students with dramatic evidence of the changes in pH surrounding each electrode. Hydroxide (OH^-) is produced around the cathode as H_2 gas is generated, creating a basic environment. The opposite is true around the anode, where the production of H^+ creates an acidic environment. The resulting color changes depend on the indicator used, as shown in the table below.

Indicator	**Neutral**	**Cathode** (where H_2 is produced)	**Anode** (where O_2 is produced)
universal	green	blue	red
bromocresol green	green	deep blue	yellow
red cabbage juice	purple	green	red

Answers to Student Questions

Step 3

No change is observed. The distilled water does not conduct electricity.

Step 4

No change is observed.

Step 5

a. Color changes will vary with the indicator used, as described in the Explanation.

b. Bubbles appear at both pencil tips (electrodes) and rise to the surface of the solution.

c. It appears that a larger volume of gas is being produced at one of the electrodes.

d. Hydrogen is being produced at the side with the larger volume of gas.

e. Water molecules contain two hydrogen atoms but only one oxygen atom.

References

Heideman, S. The Electrolysis of Water: An Improved Demonstration Procedure. *J. Chem. Educ.* **1986**, *63*, 809–10.

Kolb, K.E.; Kolb, D.K. Apparatus for Demonstrating Electrolysis on the Overhead Projector. *J. Chem. Educ.* **1986**, *63*, 517.

Investigating a Self-Inflating Balloon

Overview

In this activity, students explore the chemistry behind self-inflating balloons.

Key Concepts

- acids and bases
- carbon dioxide
- chemical reactions
- compressibility
- endothermic processes
- experimental design
- properties of gases

National Science Education Standards

Science as Inquiry Standards

Abilities Necessary to Do Scientific Inquiry

- *Students design and conduct a scientific investigation, identifying and controlling variables and making accurate measurements. (5–8, 9–12)*
- *Students ask questions about the behavior of gases based on their observations of a balloon inflated by CO_2 gas produced through a chemical reaction inside the balloon. (5–8, 9–12)*
- *Students use their observations to predict whether the product of two separate chemical reactions is the same gas (CO_2). (5–8, 9–12)*
- *Students communicate their observations of the balloon verbally and through drawings. (5–8, 9–12)*
- *Students pose testable questions and design experiments to answer these questions. (5–8, 9–12)*

Physical Science

Properties and Changes of Properties in Matter

- *Students observe that CO_2 gas is more dense than air. (5–8)*
- *In Part C, students observe that vinegar and baking soda react chemically to form a new substance (CO_2 gas) with different characteristic properties. (5–8)*

Chemical Reactions

- *Students learn that chemical reactions occur all around us by observing a reaction occurring between two common kitchen ingredients. (9–12)*

Transfer of Energy

- *Students learn that dissolving baking soda in water is an endothermic process. (5–8, 9–12)*

Structures and Properties of Matter

- *Students observe that the gas molecules generated from the reaction exert pressure on the inside walls of the balloon. (9–12)*

Science and Technology

Abilities of Technological Design

- *Students use what they learn from this investigation to identify problems for technological design (for example, designing a balloon that will not pop when excessive exterior force is applied). (5–8, 9–12)*

Part A: Student Exploration

Inflate and observe.

Materials

- self-inflating opaque and transparent balloons
- paper towel

Procedure

1. Examine the opaque balloon.

Figure 1

   a. Make some observations. *Describe what you feel inside the balloon.* Draw a picture of what you think might be inside the balloon.

   b. Activate the opaque balloon by placing it on a table and hitting the center of the balloon with the heel of your hand until something happens. Shake the balloon and listen. *Describe and record the changes you observe in the balloon.*

   c. Based on your observations as the balloon changed, revise your initial drawing of what you think is inside if needed.

2. Examine the transparent balloon.

   a. *What do you see inside it?* Compare what you see to the drawing you made. Explain how you think the toy works based on what you see and what you observed when you activated the opaque balloon.

   b. Activate the transparent balloon and see if you are right.

   c. Shake the balloon and listen. *Describe and record the changes you observe in the balloon.* Be sure to include observations you make using your senses of sight, hearing, and touch.

   d. *Compare what you observed in this balloon to what you thought made the balloon self-inflate.*

Part B: Student Exploration

Disassemble a self-inflating balloon to determine the identity of the contents.

Materials

- self-inflating balloon (transparent or opaque)
- scissors
- blue and red litmus paper
- water
- set of white powder "knowns," labeled or in original containers
  - baking soda
  - flour
  - powdered sugar
  - salt
  - citric acid powder
- paper towel
- 2 plastic cups
- set of clear liquid "knowns," labeled or in original containers
  - tap water
  - rubbing alcohol
  - vinegar
  - salt solution
  - citric acid solution made from citric acid powder

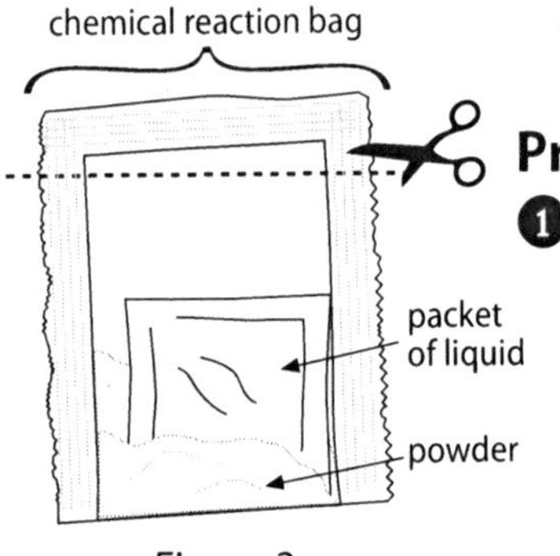

Figure 2

Procedure

1. Carefully cut open the balloon and remove the contents. Depending on the type of balloon, you may have a liquid packet plus a round, pale yellow pellet or a chemical reaction bag containing white powder and a liquid packet (Figure 2). The "unknowns" for this exploration will be the liquid (found in either type of balloon) and the solid (found in pellet or powder form, depending on balloon type). If you have the chemical reaction bag, cut off the top of the reaction bag and remove the packet of liquid. Set aside the liquid packet for now.

2. Moisten strips of blue and red litmus paper with tap water and touch each litmus strip to the pellet or push each litmus strip into the white powder as shown in Figure 3. *Record your results.* Use your observations and the acid-base testing chart to determine if the solid is acid, base, or neutral.

Acid-Base Testing Chart

Observation	Conclusion
blue strip changes to red and red strip stays red	acid
red strip changes to blue and blue strip stays blue	base
blue strip stays blue and red strip stays red	neutral

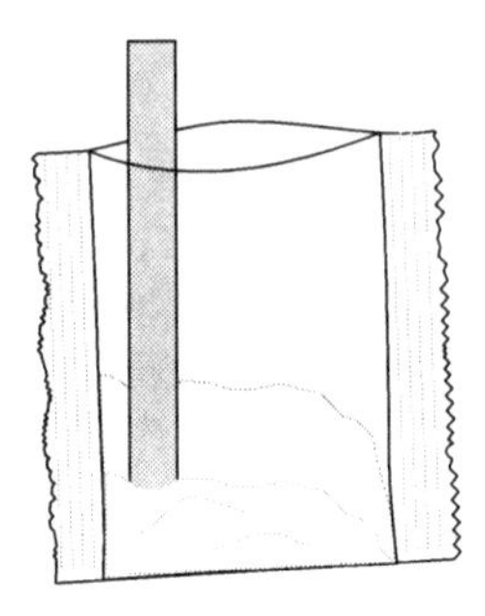
Figure 3

Figure 4

3. In a similar fashion, test the provided white powder "knowns." These are common white powders you might find around the house. *Record your observations and determine if the powders are acid, base, or neutral.*

4. Rinse off the liquid packet and dry it. Carefully cut off the corner of the packet as shown in Figure 4, and pour the liquid into a cup. Waft (wave the air above the liquid towards your nose) to determine what the liquid smells like. Test the liquid with both the red and blue litmus. (You don't have to wet the litmus before using it.) *Record your results. Is the liquid acid, base, or neutral?*

5. In a similar fashion, test the clear liquid "knowns." *Record your observations and determine if the liquids are acid, base, or neutral.*

6. Pour the pellet or powder into the cup containing the reaction bag liquid. *Make as many observations as possible.* Look. Listen. Feel the outside of the cup. Label the cup "step 6 gas" and save the cup and its contents for Part C.

7. Based on your observations and litmus tests of the known white powders and clear liquids, select one powder and one liquid that you think are the same substances as the unknown pellet (or powder) and liquid from the reaction bag. Mix the powder and liquid you selected. Observe, and compare your results with step 6. Label the cup "step 7 gas" and save the cup and its contents for Part C.

Part C: Student Exploration

So, what's the gas?

Materials

- birthday candles
- scissors
- small piece of foil
- foil cup or small foil pie pan
- graduated cylinder or measuring cup
- vinegar
- matches
- 1-L plastic soft-drink bottle
- baking soda
- teaspoon
- funnel or paper cone
- liquid packet and powder or pellet from a self-inflating balloon
- gases generated in Part B, steps 6 and 7
- Alka-Seltzer®
- unsweetened Kool-Aid®

Procedure

 *If the candle burns too low to light comfortably, put in a new one. Keep flammable materials such as clothing, hair, and papers away from the candle flame.*

1. Cut a birthday candle so that the top of the wick is below the rim of the foil cup or pie pan. Use a bit of foil to make a holder so the candle will stand up. Make sure the wick is still below the rim, and adjust as needed.

2. Pour about 80 mL (1/3 cup) vinegar into the 1-L bottle. Although the bottle contains a little vinegar, it is mostly filled with the gases that make up air. Now you are going to try to put out a candle by "pouring" the gas (not the liquid) out of the bottle.

    a. Light the birthday candle. Tip the bottle as though you are pouring the bottle's gas into the foil cup. (See Figure 5.) Stop before any liquid comes out. *What happens?*

    b. Let a few drops of vinegar fall into the cup (but not on the flame). *What happens now?*

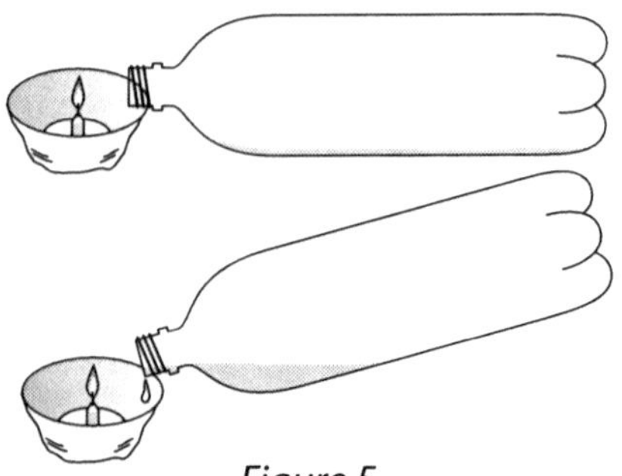

Figure 5

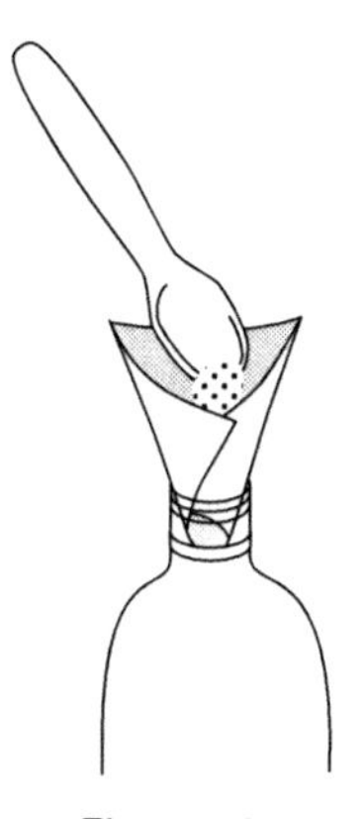
Figure 6

c. Pour 5 mL (1 teaspoon) baking soda into the bottle containing the vinegar. (You may want to use a funnel or paper cone to help you pour. See Figure 6.) *What do you observe?*

d. Swirl the bottle to make sure the liquid and solid are mixed. When the fizzing dies down, try pouring the gas into the foil cup with the lighted birthday candle, just as you did before. *What happens?* (You may need to keep pouring until a few drops of liquid fall into the foil cup, but don't pour the liquid onto the burning candle.)

3. Remember the gas produced when you mixed the liquid packet and powder or pellet in Part B, step 6? *What do you think will happen when you pour that gas onto a lighted candle? Try it and see.*

4. Pour the gas you made in the cup from Part B, step 7, onto a lighted candle. *Observe and record your results.*

5. To collect additional information, test the gas generated by the recipes below on a lighted candle. Remember to swirl the bottle to mix the ingredients well. Rinse the bottle out with water between tests.

- recipe A: 1 packet unsweetened Kool-Aid, 5 mL (1 teaspoon) baking soda, and about 80 mL (⅓ cup) water
- recipe B: 2 Alka-Seltzer tablets (broken into pieces) and about 80 mL (⅓ cup) water

Identify the solid and liquid used in the balloon. What is your evidence?

Instructor Notes

Tips and Instructional Strategies

- Don't show students the transparent balloon until they've completed Part A, step 1. If you don't have transparent balloons, you may want to cut open an opaque balloon to show students the contents. The self-inflating balloons come in two styles—one style contains a round pellet and a liquid packet. The other style contains a chemical reaction bag which holds powder and a liquid packet. Either style is fine for this lesson because both types of balloon contain the same substances and produce identical reactions.

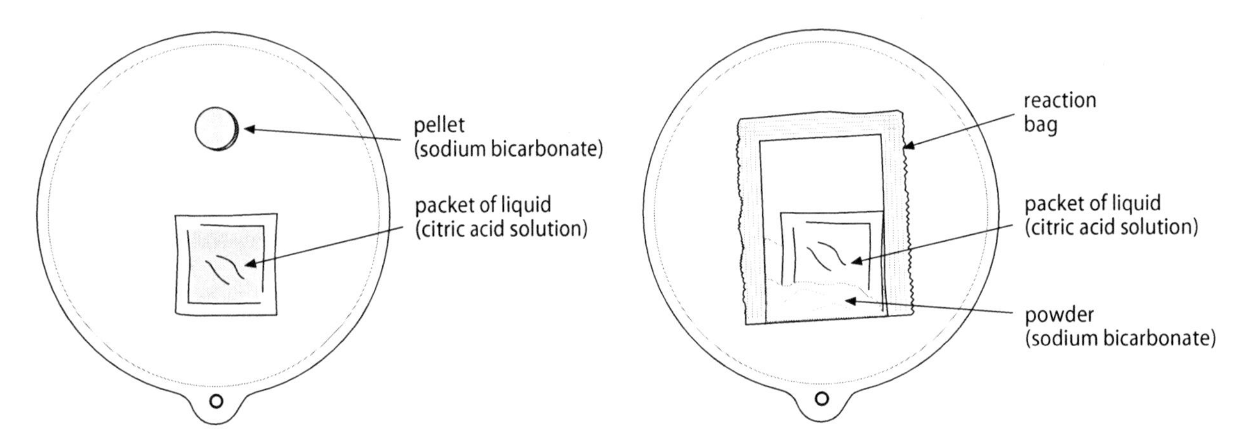

- For Part B, if litmus paper is not available, you can substitute coffee filter paper soaked in red cabbage indicator, dried and cut into strips. The colors of this indicator are as follows: base–green, acid–red, and neutral–purple. The indicator can be prepared by one of these methods:
  - Put approximately ¼ head of red cabbage into a blender or food processor. Add a little water and blend the cabbage into a slurry (1–3 minutes). Pour the slurry through a strainer, collecting the juice in a beaker.
  - Chop ¼ head of red cabbage and put it into a beaker. Add rubbing alcohol to cover the cabbage. Stir periodically. After about an hour, decant the liquid and discard the cabbage pieces.
- For Part B, citric acid powder is sold in grocery stores as Fruit Fresh.
- For Part C, making small cups from aluminum foil (as described below) can help students be more successful in extinguishing the candle by providing a smaller volume to fill with CO_2 gas. A small 15-cm (about 6-inch) diameter aluminum pie plate is a good alternative. In either case, make sure that the candle wick is below the rim of the cup or pan to ensure that the wick gets covered by the CO_2 gas as the students pour the gas.

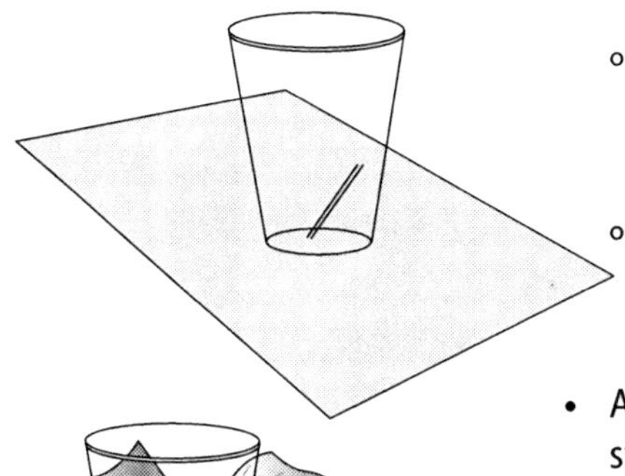

Foil cup

- Place a cup in the middle of a piece of foil. Make sure you have at least 5 cm (2 inches) of foil on all sides of the cup. Wrap the foil around the cup to make a foil cup. (See figure.)
- Remove the cup, and measure the height of the foil cup. Make the foil cup about 5 cm (2 inches) tall by folding down the edges if needed. If students are using this height of cup, a 4-cm (about 1½-inch) tall candle will be ideal.

- After Part C, step 2c, have students predict the identity of the gas produced. In step 2d, the candle should go out. This is evidence that the gas is CO_2. If some students' candles did not go out, have them add more vinegar and baking soda and try again. They may need a little practice aiming this colorless, odorless gas so it fills the foil cup rather than spilling elsewhere.
- At the end of Part C, you may want students to experiment with other recipes to produce CO_2. Have them try various combinations using an acid (vinegar or undiluted lemon, lime, or orange juice) and a source of carbonate or bicarbonate ions (pieces of eggshell, some brands of chalk, seashells, baking powder, or limestone rock). The reactions with these solids are typically much slower and the amount of gas produced can be much less than in the reactions students have already seen, so instruct them to be patient and very observant. You can also have students design their own self-inflating balloon.
- Challenge students to pose testable questions based on what they've done and learned in the lesson. They can design experiments and collect evidence to answer these questions, formulate claims about their findings, and, if time allows, present and defend their claims with their group or class. While we strongly recommend that students develop their own testable questions, you may need to seed the discussion with possible questions such as these: What effect does the temperature of the balloon have on the self-inflating action? (That is, what would be different if students used a self-inflating balloon that was hotter or colder than room temperature?) How much weight can you place on the inflated balloon before it pops? In Part B, steps 6 and 7, why was it unnecessary to cover the cup of CO_2 once the gas was made?

Explanation

As students discover in Part A, the citric acid solution is separated from the sodium bicarbonate pellet or powder until you hit the balloon and break the liquid packet. Shaking the balloon helps the chemicals mix so they can react and produce more gas. In the type of balloon with a reaction bag, the gas produced increases pressure in the reaction bag until the bag bursts. In both types of balloons the gas fills the balloon. Once the balloon is filled, additional gas formation exerts more and more pressure on the inside walls. The opaque balloon is made of strong material and is also securely sealed so that it can withstand the pressure without bursting. In fact, it's usually strong enough to hold up a person who weighs under 113 kg (about 250 pounds).

In Part B, students take a closer look at the solid and liquid ingredients outside of the balloon and relate the reaction they observe to the inflation of the balloon. They discover that the dissolving of baking soda in water or other aqueous solutions is an endothermic process. Since the citric acid solution has virtually no odor, students can typically rule out vinegar as the unknown liquid when they waft the sample.

In Part C, students further investigate the gas responsible for inflating the balloon—carbon dioxide (CO_2). Students demonstrate the presence of invisible CO_2 gas by using it to displace the oxygen a candle flame needs to burn. CO_2 is a nonflammable gas. Because CO_2 is more dense than air, it can be poured downward, displacing the less dense air and collecting in the aluminum foil cup where the candle sits. As CO_2 collects in the cup, the air in the cup is pushed up and out by the more dense CO_2 gas. As CO_2 collects around the flame, the flame loses its oxygen supply and goes out.

The recipes in Part C and in the Tips and Instructional Strategies produce CO_2 because each includes an acid and a source of carbonate or bicarbonate ions. In Kool-Aid, citric acid is one of the powdered ingredients. Alka-Seltzer is a mixture of solids, including sodium bicarbonate, acetylsalicylic acid (aspirin), and citric acid. Eggshells, seashells, limestone rock, and some brands of chalk all contain calcium carbonate.

acid	+	carbonate or bicarbonate ions	=	carbon dioxide gas	+	other stuff
citric acid; vinegar; lemon, orange, or lime juice		*baking soda, eggshell, seashell, limestone rock, chalk, or baking powder*				

Answers to Student Questions

Part A

Step 1

a. *The balloon contains a rectangular object. The object feels lumpy and a little squishy. (The balloon may also contain a round pellet.)*

b. *Something inside flattens or breaks. A small bulge starts to form, and then there's a hissing sound. (In one style of balloon, a loud pop may be observed.)*

Step 2

a. *The opaque balloon either contains a packet of clear liquid and a pale yellow round pellet or a little bag, which contains white powder and a packet of liquid. (See illustrations in Tips and Instructional Strategies.)*

b. *The liquid and the yellow pellet (or white powder) start to bubble and make hissing noises as soon as the liquid bursts out of its packet. (If using the type of balloon with a reaction bag, the reaction bag starts to inflate and soon pops open with a loud sound.) The liquid keeps bubbling and hissing more and more, and the balloon starts to inflate and get cold. After a while, the balloon is very firm, and the bubbling and hissing slow down, then stop.*

Part B

Step 2

The red strip changes to blue. The blue strip stays blue. The pellet (or *powder) is a base.*

Step 3

a. Baking soda: The red strip changes to blue. The blue strip stays blue. Powder is a base.

b. Flour: No change, neutral.

c. Powdered sugar: No change, neutral.

d. Salt: No change, neutral.

e. Citric acid: The blue strip changes to red. The red strip stays red. Powder is an acid.

Step 4

The blue strip changes to red. The red strip stays red. Liquid is an acid.

Step 5

a. Tap water: No change, neutral or slightly acidic or basic. (Water can vary in pH due to the presence of dissolved minerals from the ground or CO_2 *from the atmosphere.)*

b. Rubbing alcohol: No change, neutral.

c. Vinegar: The blue strip changes to red. The red strip stays red. Liquid is an acid.

d. Salt solution: No change, neutral.

e. Citric acid solution: The blue strip changes to red. The red strip stays red. Liquid is an acid.

Step 6

The mixture of the reaction bag ingredients bubbled, hissed, and the container got cold, just like in the balloon.

Part C

Step 2

a. The candle does not go out.

b. The candle still does not go out.

c. The mixture reacts, forming a gas.

d. After pouring, the candle goes out.

Step 3

The gas will put out the candle.

Step 4

The gas will put out the candle.

Step 5

The solid and liquid are sodium bicarbonate and citric acid, based on the observed properties and tests performed.

Alka-Seltzer Poppers

Overview

Students investigate the effects of changing the volume and temperature of the water used to facilitate the reaction of Alka-Seltzer®.

Key Concepts

- chemical reactions
- experimental design
- gases
- kinetics
- solubility

National Science Education Standards

Science as Inquiry

Abilities Necessary to Do Scientific Inquiry

- *Students use evidence and logical argument gathered from the experiments to develop explanations and predictions. (5–8, 9–12)*
- *Students gain experience in identifying and controlling variables as they plan and conduct experiments to determine the impact of changing the volume and temperature of water on a chemical reaction and to determine the identity of the gas produced in the reaction. (5–8, 9–12)*

Physical Science

Properties and Changes of Properties in Matter

- *Students investigate properties of the carbon dioxide gas (CO_2), including the solubility of CO_2 in water at different temperatures. (5–8)*
- *Students observe chemical reactions in progress and discover that components of Alka-Seltzer react to produce new substances with different characteristic properties. (5–8)*

Chemical Reactions

- *Students investigate the acid-base neutralization reactions between a base (sodium bicarbonate) and each of two weak acids (acetylsalicylic acid and citric acid). These three reactants begin as a mixture of solids in Alka-Seltzer tablets. When Alka-Seltzer is placed in water, the water dissolves the reactants and facilitates the reactions. The most obvious product of the reactions is CO_2 gas. (9–12)*
- *By changing one variable in each trial and timing how long it takes to pop the lid off a film canister, students are able to analyze how each variable changes the reaction rate. (9–12)*

Motions and Forces

- *Students observe the motion of the lid being pushed off the canister by the force created from the buildup of CO_2 gas. (5–8, 9–12)*
- *With the lid on the film canister, the gas in the canister becomes compressed as CO_2 is produced during the chemical reaction. This compression causes an increase of pressure in the canister, which eventually causes the lid to pop off the canister. (9–12)*

Student Exploration

Let's investigate the addition of Alka-Seltzer to water.

Materials

- 5 or more Alka-Seltzer tablets
- clear container such as a plastic cup or glass
- room-temperature water
- 35-mm film canister with lid
- teaspoon measure
- tablespoon measure
- ice-cold water
- very hot tap water
- other materials as needed for the student-designed experiments
- (optional) 60-mL (¼-cup) measure

Procedure

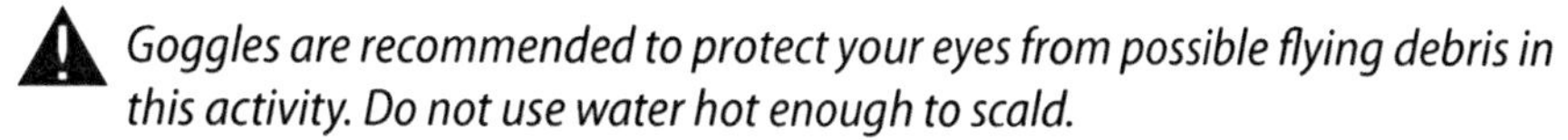

⚠ *Goggles are recommended to protect your eyes from possible flying debris in this activity. Do not use water hot enough to scald.*

1. Pour about 60 mL (¼ cup) room-temperature water into a clear container. Add an Alka-Seltzer tablet to the water and observe. Keep the solution for use in step 2. *Describe what happens. Determine and record the active ingredients of Alka-Seltzer tablets (as shown on the packaging). What role does water play in this system?*

2. Place 15 mL (1 tablespoon) of the solution from step 1 in the film canister. Add half an Alka-Seltzer tablet to the canister and immediately put on the lid. Put the canister on a table and hold it in place, away from your face. *What happens? Explain your observations.*

3. Design and perform an investigation to determine if the observations in step 2 are reproducible. *What did you do? What did you find out?*

4. Repeat step 2 with 15 mL (1 tablespoon) room-temperature water that has not had an Alka-Seltzer tablet dissolved in it already. *How do the results compare with those in step 2?*

5. Rinse the canister and place 5 mL (1 teaspoon) room-temperature water in it. Add half an Alka-Seltzer tablet to it and immediately put on the lid as before. *Compare your observations with those in step 4 and explain any similarities and differences. What do you think would happen if you added 30 mL (2 tablespoons) water to the canister before adding the Alka-Seltzer?* Try this variation using the same technique. *Explain your observations.*

6. Design an experiment to determine the impact of changes in the water temperature on the system. Do not use water hotter than hot tap water. Describe what you did and observed. Explain your observations.
7. Devise an experiment to determine the identity of the gas produced by the Alka-Seltzer in water.

Instructor Notes

Tips and Instructional Strategies

- Upon completion of the activity, be sure that students have discovered that the water acts as a solvent to dissolve the active ingredients (sodium bicarbonate and the two acids) so that they can react. The time it takes for the lid to pop off is related to the temperature of the system and to the volume of the water used.
- Have students do web research to determine the solubility of CO_2 in water and use this information to explain experimental results.
- For older grades in which the discussion of acid-base systems includes the Brønsted–Lowry concept, you will want to point out that sodium bicarbonate is a proton acceptor (Brønsted–Lowry base) in this system. Also, make sure that students understand that two different acids are present (citric acid and acetylsalicylic acid) and thus two different acid-base neutralization reactions need to be considered. These reactions are described in the Explanation.
- For lower-grade students who may not know about the Brønsted–Lowry definitions of acids and bases, the reactions that occur can be looked at as two double-replacement reactions that form carbonic acid (H_2CO_3):

  $$3\,NaHCO_3(aq) + H_3C_6H_5O_7(aq) \rightarrow 3\,H_2CO_3(aq) + Na_3C_6H_5O_7(aq)$$

  $$NaHCO_3(aq) + HC_9H_7O_4(aq) \rightarrow H_2CO_3(aq) + NaC_9H_7O_4(aq)$$

  with the carbonic acid in both cases decomposing into water and CO_2 gas:

  $$H_2CO_3(aq) \rightarrow H_2O(l) + CO_2(g)$$

  Removing $CO_2(g)$ shifts the equilibrium to produce more $CO_2(g)$. This lowers the hydrogen ion concentration, so the pH increases. The amount of H_2CO_3 actually present is small. Shakhashiri reminds demonstrators that the ratio of CO_2 to H_2CO_3 is about 600 to 1 at 25°C.

Explanation

The active ingredients in an Alka-Seltzer tablet are sodium bicarbonate ($NaHCO_3$), citric acid ($H_3C_6H_5O_7$), and aspirin (acetylsalicylic acid, $HC_9H_7O_4$). When the solid reactants are dry, no reaction occurs. When water is added and the reactants dissolve, the basic sodium bicarbonate reacts with the citric and acetylsalicylic acids by acid-base neutralization reactions. One product is gaseous carbon dioxide (CO_2), which is responsible for the bubbling that occurs. Alka-Seltzer is sealed in a foil wrapper to keep it from contacting moisture in the air and reacting prematurely.

The two reactions occurring with citric acid and aspirin, respectively, are:

$$3\ NaHCO_3(aq) + H_3C_6H_5O_7(aq) \rightarrow 3\ H_2O(l) + 3\ CO_2(g) + Na_3C_6H_5O_7(aq)$$

$$NaHCO_3(aq) + HC_9H_7O_4(aq) \rightarrow H_2O(l) + CO_2(g) + NaC_9H_7O_4(aq)$$

The film canister has a fixed volume. When the Alka-Seltzer tablet is added, the gas generated is trapped, so the pressure increases. When sufficient CO_2 has been generated, the pressure is large enough to cause the lid to pop off.

Several factors influence the amount of time it takes for the lid to pop off. The amount of water in the canister is one factor. With a larger volume of water, the volume of air initially present is less. Since the volume available for the gas is less, less time is needed to generate sufficient CO_2 to pop the lid off. Likewise, the larger the piece of Alka-Seltzer, the faster the reaction, since the concentrations of the reactants is higher.

Carbon dioxide is somewhat soluble in water. With a larger volume of water, more CO_2 has to be generated before the water is saturated with it. When water already saturated with carbon dioxide is used, less time is needed for the lid to pop off since none of the CO_2 generated dissolves. But if tap water is used, some of the carbon dioxide generated dissolves in the water and more time is needed for the lid to pop off at the same temperature.

The temperature of the water also affects the rate of the reaction. First, the higher the temperature, the faster the rate of reaction, so it takes less time for the lid to pop off. A second factor is that the higher the temperature, the less soluble carbon dioxide is in water. Both these factors then predict that a shorter time will occur with hot water and a longer time with cold water, assuming the same amounts of water and Alka-Seltzer are used.

Answers to Student Questions

Step 1

a. *The solution bubbles and the Alka-Seltzer tablet disintegrates.*
b. *The active ingredients in Alka-Seltzer are sodium bicarbonate, citric acid, and aspirin (acetylsalicylic acid).*
c. *The water dissolves the active ingredients and facilitates the reaction.*

Step 2

The film canister lid pops off after a few seconds. As the gas is produced, pressure builds up inside the canister until it is great enough to pop the lid off.

Step 3

Students should repeat the experiment several times to check for reproducibility. They should measure the time it took for the lid to pop off for comparison. Or they could do several identical, simultaneous runs (team work) to see if the lids pop off at the same time. They should find that the results are reproducible to within a second or two.

Step 4

The top should take a little longer to pop off the film canister. In step 2, the water was already saturated with carbon dioxide, so all the gas produced went toward increasing the pressure within the canister. In step 4, some of the carbon dioxide gas produced dissolved in the water, so it took longer to generate enough gas to cause the top to pop.

Step 5

a. *With 5 mL (1 teaspoon) water, it takes longer for the top to pop off the canister. There are several factors to consider. There is less water so the tablet may dissolve less rapidly and less of the carbon dioxide dissolves in the water. There is more air space so it takes longer to produce enough gas to increase the pressure enough to pop the top.*

b. *With 30 mL (2 tablespoons) water, the canister is nearly filled. There is more water for the tablet and the carbon dioxide to dissolve in. There is very little space for gases so a relatively small amount of gas will get the pressure high enough to force the top off.*

References

Sarquis, A.; Woodward, M. Alka Seltzer Poppers: An Interactive Exploration, *J. Chem. Educ.* **1999,** *76* (3), 385–386.

Shakhashiri, B.Z. *Chemical Demonstrations,* Vol. 2; University of Wisconsin: Madison, WI, 1985; pp 114–120.

Penny Sandwiches

Overview

This lesson allows students to remove the inside of the penny, leaving only the thin copper coating.

Key Concepts

- activity series
- oxidation-reduction reaction (redox)

National Science Education Standards

Science as Inquiry

Abilities Necessary to Do Scientific Inquiry

- *Students use logic and evidence to formulate explanations about what they observe over time when pennies are soaked in an acid. (5–8, 9–12)*

Physical Science

Properties and Changes of Properties in Matter

- *Students conclude that a chemical reaction occurs when they observe bubbles as the zinc in the penny reacts with an acid. (5–8)*

Chemical Reactions

- *Students learn that an oxidation-reduction reaction occurs when zinc in the penny reacts with the acid. (9–12)*
- *Older students conclude that the bubbles they observe during this single replacement reaction are hydrogen gas. The reaction is complete when the bubbles stop forming. (9–12)*
- *In the variation, students observe pennies in different acids over the course of a week and realize that chemical reactions can occur at different rates, depending on factors such as acid concentration. The redox reaction of zinc with acid proceeds at a faster rate in more highly concentrated acid. (9–12)*

Science and Technology

Understanding about Science and Technology

- *Students learn why pennies made after 1983 need a protective copper coating. (5–8, 9–12)*

Student Exploration

Can you dissolve a penny?

Materials

- penny (dated 1983 or later)
- file or coarse sandpaper
- 6 M hydrochloric acid (HCl)
- clear wide-mouthed container (with appropriate loose cover)

➤ *You will need to cover, not seal, the container. The pressure of the gas generated could be sufficient to shatter a sealed container.*

Procedure

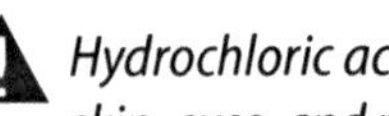

Hydrochloric acid is corrosive and can cause burns. The vapor is irritating to the skin, eyes, and respiratory system. Eye protection is required. Do this activity only in a well ventilated area. Use a watch glass or other appropriate cover.

1. Using the file or coarse sandpaper, completely remove the copper coating from the edge of the penny. Position the penny in the container so that it is standing on its edge against the side of the container. Pour the 6 M hydrochloric acid (HCl) into the container. Observe the penny for 4 or 5 minutes. *What do you see? What is occurring?*

2. Observe the penny over the course of 1–2 days and describe any differences in appearance over time. *What specific process do you think is occurring? How do you know when the process is complete?*

Instructor Notes

Tips and Instructional Strategies

- If you prefer not to use 6 M hydrochloric acid (HCl), you can substitute ½ cup vinegar, lemon juice, or a less concentrated HCl solution. With these alternatives, the reaction can take more than a week to complete and the vinegar or lemon juice will need to be replaced as it dries out. As a variation, you can split the class into groups and have each try a different solution and compare their results.
- It may be helpful to periodically stir the acidic solution.
- For lower grades, you might want to ask students whether the activity shows a physical or a chemical change. Ask them what their evidence is.
- For advanced classes, ask students if they can identify the gas forming the bubbles in step 1. At the end of the activity, have students write an equation representing the processes they observed. Have them use their observations to rank the elements hydrogen, copper, and zinc from most to least reactive. (See Explanation.)
- From 1864 until 1982, the usual composition of a penny (the coin's actual name is "cent") was 95% copper and 5% zinc and tin. Because of the increasing cost of copper, pennies are now composed of a zinc interior with a thin copper coating comprising only 2.4% of the penny.
- Challenge students to explain the reason for coating the penny with copper instead of making it entirely out of zinc.

Explanation

Pennies minted since 1983 are composed of a zinc core with a copper coating. This experiment illustrates the relative activity towards oxidation for zinc (Zn), hydrogen (H_2), and copper (Cu) and illustrates why the Zn core of the penny requires the protective copper coating. Once the edge of the penny has been filed away, both metals are exposed to the hydrochloric acid (HCl).

The activity series of metals lists metals and H_2 in order of ease of oxidation. The activity series is

$$Zn > H_2 > Cu$$

This means that Zn is more easily oxidized than either H_2 or Cu. In this reaction, the oxidizing agent (the species being reduced) is hydrogen ion (H^+) from the acid solution. According to the activity series, Zn can be oxidized in an acid solution, but Cu can't, since it falls below H_2 on the activity series.

The net equation is

$$Zn(s) + 2H^+(aq) \rightarrow Zn^{2+}(aq) + H_2(g)$$

The Zn is oxidized to H^+, and the H^+ is reduced to H_2. Thus the bubbles formed are H_2. Since Cu is below H_2 on the activity series, it does not react in the acid solution. As long as H^+ is present, the reaction will continue until all of the Zn has reacted, leaving only the Cu foil that coated the penny. The 6 M HCl has a much higher H^+ concentration than the other solutions suggested in Tips and Instructional Strategies and the reaction occurs rapidly.

Answers to Student Questions

Step 1

a. Bubbles of gas are visible.

b. The formation of a gas indicates a chemical reaction.

Step 2

a. The reaction of zinc with an acid is occurring.

b. The reaction is complete when bubbles stop forming.

References

Miller, J.M. Analysis of 1982 Pennies. *J. Chem. Educ.* **1983,** *60*, 142.

Yoeman, R.S. *A Guide Book of United States Coins,* 24th ed.; Western Publishing Company: Racine, WI, 1970.

Things That Glow

Overview

This engaging lesson illustrates the differences between fluorescent and phosphorescent materials. Students are shown how electrons in different materials respond differently when exposed to light. They investigate variables that affect the brightness and duration of glow in activated zinc sulfide. Students also see, using colored LEDs, how differing wavelengths of light have different energies.

Key Concepts

- electromagnetic spectrum
- electrons
- energy levels
- excited state
- experimental design
- fluorescence
- ground state
- light
- models
- phosphorescence
- photons
- quantum of energy
- wavelength

National Science Education Standards

Science as Inquiry Standards

Abilities Necessary to Do Scientific Inquiry

- *Students are shown a kinesthetic model of fluorescence and are asked to differentiate between the process described by the model and what occurs in real atoms. (5–8, 9–12)*
- *Students use a kinesthetic model of fluorescence to understand how the photons of fluorescent materials respond to ultraviolet light. (5–8, 9–12)*
- *Students conduct an investigation with a phosphorescent material. (5–8, 9–12)*
- *Students use their observations to develop explanations about the intensity of glow-in-the-dark materials. (5–8, 9–12)*
- *Students pose testable questions and design experiments to answer these questions. (5–8, 9–12)*

Physical Science

Properties and Changes of Properties in Matter

- *Students learn that fluorescence and phosphorescence are characteristic properties of some materials. (5–8)*

Transfer of Energy

- *Students observe that phosphorescence and fluorescence occur when matter interacts with electromagnetic radiation. (5–8)*
- *Students observe that light energy can be stored by some chemicals (like activated zinc sulfide) and released over time. (5–8)*

Structure of Atoms

- *Students learn that the behavior of electrons accounts for some characteristic properties of matter. (5–8)*

Interactions of Energy and Matter

- *Students observe that phosphorescent and fluorescent objects are "excited" by electromagnetic radiation and emit visible light. (9–12)*
- *Students learn that atoms or molecules gain or lose energy only in particular discrete amounts and absorb or emit light at wavelengths corresponding to this energy. (9–12)*

Part A: Teacher Demonstration

Jump to a higher state! What is fluorescence?

Materials

- fluorescent items such as the following:
  - day-glow paints and papers
  - laundry detergent with optical brighteners and labels from these products
  - freshly washed, white clothing
  - minerals that fluoresce (such as opal, fluorite, calcite, willemite, or sheelite)
  - petroleum jelly
  - quinine water (tonic water)
  - Vaseline glass (a type of antique glass the color of Vaseline® petroleum jelly)
- UV light source
- 4–5 small balls or similar non-breakable items
- flashlight
- desk or table

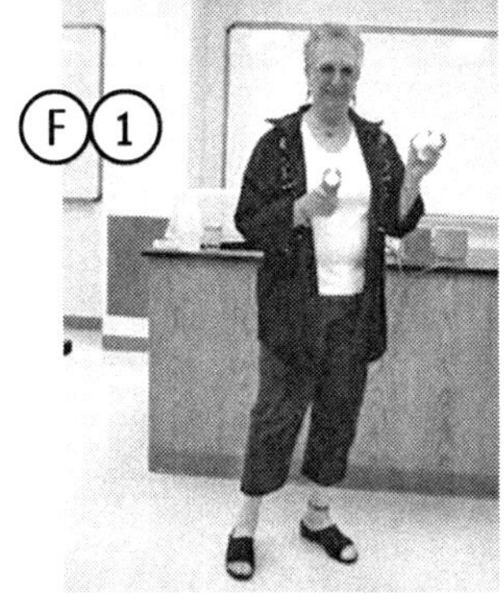
F1

Procedure

1. Without revealing their special property, show the class a group of items that will fluoresce under ultraviolet (UV) light. Ask students what they think these items have in common.

2. Darken the room. Shine a UV light on the items and ask students what they observe. Turn the UV light off, and ask students what they observe now. Ask the students again what they think these items have in common.

⚠ *The UV light emitted from a black light or other UV light source can cause severe eye damage. Use carefully so that no one can look directly at the light.*

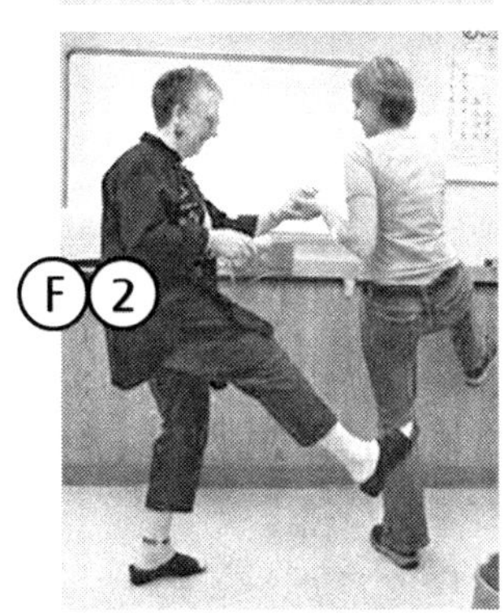
F2

3. Tell students that this phenomenon is called fluorescence. Explain fluorescence using the following kinesthetic model (See photos F1–F4):

   a. Have a student volunteer stand at the front of the room to represent one of the electrons in the fluorescent material. This electron is in its ground state.

   b. You will play the role of the UV light: Hold the balls (to represent photons) in one hand and a flashlight in the other hand.

   c. Shine the flashlight on the volunteer and transfer the photons (balls) to the student volunteer, as you give the student a pretend kick to bump him or her up to a higher energy level, which is represented by the student moving up on a desk, table, stairs, or ladder. Explain that the UV light provides energy to "kick" electrons up to an excited (higher) state.

F3

d. Have the student move back down to the ground level, throwing the balls into the air in the process. Explain to the class that when electrons return to ground state, electromagnetic radiation is given off, sometimes as visible light. (Photo F4 shows the student "electron" in the process of moving to ground state.) Explain that fluorescence involves the object emitting light almost immediately after absorbing it. When the light source is removed, the object immediately ceases to glow.

4. Point out that this kinesthetic model does not perfectly reflect the nature of real electrons. Like all models, it is impossible to reproduce or replicate all aspects of the real phenomenon. Ask students to list some of the limitations of this model. (See Instructor Notes for a discussion of the model's limitations.)

P1

P2

Repeat the steps shown previously in photos F1–3. From the higher energy state shown in photo F3, the student "electron" drops to a lower but still excited intermediate level and gives off energy, but not in the visible range (photo P1). The student "electron" then drops to the ground state, giving off energy in the visible range (photo P2).

Part B: Teacher Demonstration

How is phosphorescence different?

Materials

- sheet of phosphorescent vinyl or other phosphorescent material (such as stars, bandages, toys, or paint)

Procedure

1. Before beginning, be sure that the glow-in-the-dark vinyl has not been exposed to light recently and is oriented face down.
2. Sandwich the vinyl between your hands, continuing to keep it face down. (If using a smaller phosphorescent item, block a part of it from the light using an alternative method.)
3. Keeping your hands in place, flip the vinyl face up so it is exposed to the light in the room for 10–15 seconds. (The purpose of holding your hand in place on the vinyl is to show a dark handprint on the glowing vinyl sheet.)
4. Turn off the room light, remove your hand, and hold the vinyl up for all to see. Discuss the observations.
5. Tell students that this phenomenon is called phosphorescence. Ask the class how their observations of fluorescence in Part A differ from their observations of phosphorescence. Have students work in groups to propose models to explain phosphorescence. Allow groups to share the models with the class and encourage discussion of the usefulness and limitations of each model.
6. If students have not already proposed the explanation, explain that like fluorescence, phosphorescence involves the loss of energy of excited state electrons. With a student volunteer, demonstrate the kinesthetic model of phosphorescence shown at left. (See photos P1–P2 and caption.) Ask students to watch carefully and try to figure how out this demonstration differs from the fluorescence kinesthetic model.

Part C: Student Exploration

Getting glue to glow.

Materials

- zinc sulfide, copper-doped powder (glow-in-the-dark pigment)
- white school glue
- small paper plates
- measuring spoons
- cotton-tipped swabs or craft sticks
- dark paper

Procedure

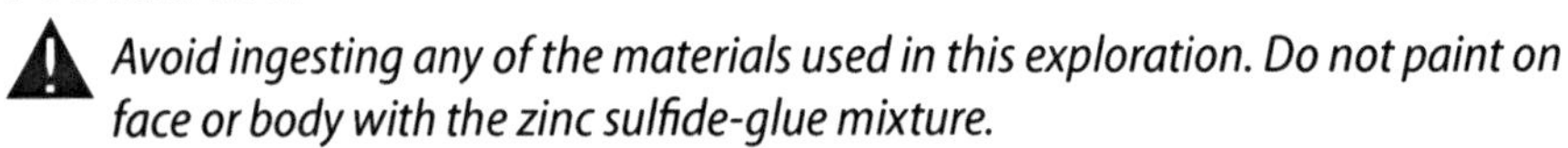

Avoid ingesting any of the materials used in this exploration. Do not paint on face or body with the zinc sulfide-glue mixture.

1. Pour 1.25 mL (¼ teaspoon) activated zinc sulfide powder in a pile on a paper plate.
2. Pour 5 mL (1 teaspoon) glue next to the activated zinc sulfide powder. Do not mix them yet.
3. Turn the room lights off and view your samples. *Describe what you see.*
4. Turn the room lights back on, and use the swab or craft stick to mix the glue and zinc sulfide. Use this mixture to paint a design on a piece of dark paper. Turn the room lights off and view.

Part D: Teacher Demonstration

Do different colors of light have the same energy?

Materials

- set of red, white, green, and blue LEDs
- sheet of phosphorescent vinyl or other phosphorescent material

Procedure

1. Before beginning, be sure that the phosphorescent sample being used has not been exposed to light recently and is oriented face down.
2. With the room as dark as possible, use the lighted tip of the white LED to write on the phosphorescent sample. Introduce or review the idea that white (visible) light is made up of the familiar spectrum of colors from violet to red.
3. Ask the students to predict what they would observe if you wrote with the red LED, and then do so. Have them note the color, intensity (as none, low, medium, or high), and how long any observed light persisted.
4. Repeat step 2 with green and then blue lights.
5. Ask students to make a claim (a statement they feel is true based on evidence) about the energy associated with different colors (wavelengths) of light. Be sure they state the evidence for the claim. Ask them to predict what would happen if the phosphorescent sample were exposed to other colors of LEDs, like orange, yellow, violet, and ultraviolet.
6. Use the observations to initiate a discussion of the energy associated with the different wavelengths of light in the visible spectrum and how the energy (E) of a photon is directly proportional to the frequency (ν) of the light by the relationship: $E = h\nu$; where h is Planck's constant.

Instructor Notes

Tips and Instructional Strategies

- In Part A, step 4, help students to see the following limitations of the kinesthetic model:
    - UV light, while a type of electromagnetic radiation, is not a type of visible light. Therefore, the use of the flashlight is not accurate.
    - Our model includes only one electron in a fluorescent material that is affected by electromagnetic radiation, while an actual fluorescent material would contain countless numbers of these.
    - Our model shows one excited state and one stabilized excited state. Usually several excited states are possible.
- Part B can be done as a student exploration if you have enough phosphorescent items to share.
- For Parts B–D, turning off the lights or darkening the room as much as possible will maximize the impact of the phosphorescence. Once the sheet has been exposed to room light, it will need to be placed in the dark a few minutes before being used again.
- For Part D, use LEDs of similar size and shape to assure that factors such as the light intensity and cone angles are approximately similar. Even though similar LED devices may have similar batteries or circuits, the energy required to light the different colored diodes is different since each diode is a different material, requiring a different excitation energy and emitting light of a different energy.
- Since no phosphorescence is observed with the red LED, the importance of describing "no effect" may be difficult for students to understand without having seen some effect in the blue and to a smaller extent green LEDs. You may want to show them the red LED again after Part D, step 4. Green LEDs vary somewhat, and you may want to test yours in advance to make sure it produces at least weak phosphorescence. If the green LEDs do not glow to a small extent, the order of increasing energy of the three different color of light, red<green<blue, will be difficult for students to determine.
- Challenge students to pose testable questions based on what they've done and learned in the lesson. They can design experiments and collect evidence to answer these questions, formulate claims about their findings, and, if time allows, present and defend their claims with their group or class. While we strongly recommend that students develop their own testable questions, you may need to seed the discussion with possible questions such as these: How do each of the following variables affect the light emitted by fluorescent/phosphorescent materials: duration of exposure; light source used (incandescent, fluorescent, sunlight, candle light); distance between the light source and the fluorescent/phosphorescent materials; and temperature of the fluorescent/phosphorescent materials?

If students investigate the effect of temperature, they should find that temperature affects phosphorescence but not fluorescence. To explain their observations, students will need to understand excitation and emission. Temperature has no observable effect on excitation or emission in fluorescence or excitation in phosphorescence. However, emission from the stabilized excited state, as in phosphorescence (represented by the intermediate level in Part B) depends on the thermal energy available. The phosphorescent vinyl glows for a shorter period in warm water and a longer period in ice water.

Explanation

Visible light and ultraviolet (UV) light are two types of electromagnetic radiation that play important roles in fluorescence and phosphorescence. Electromagnetic radiation surrounds us everywhere, day and night. It consists of waves of electric and magnetic fields that travel through space at the speed of light. Depending on the wavelength, we see electromagnetic radiation as visible light, feel its presence as infrared heat, receive it through a radio as sound, or use it to cook our food in a microwave oven. (See figure.)

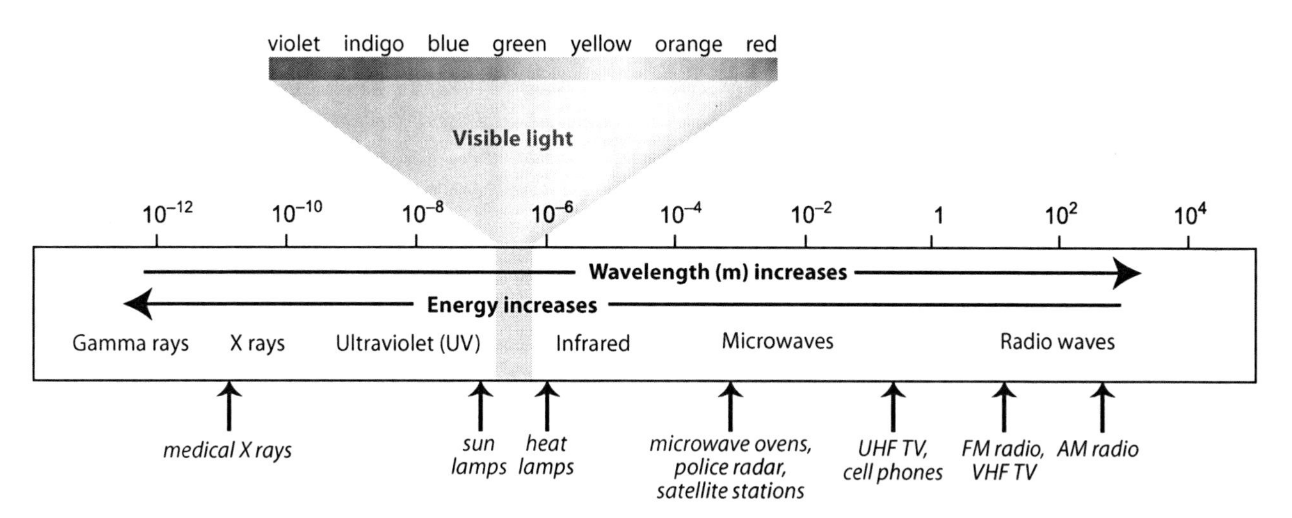

Visible light is made up of the familiar spectrum of colors from violet to red. UV is the more energetic form of light just past violet in the electromagnetic spectrum. Humans cannot sense UV radiation until it has done its damage and we feel its effect as sunburn. Infrared radiation can be sensed as heat but also lies outside the visible range of the spectrum. Devices sensitive to infrared radiation are used to check for areas of heat loss in houses with the goal of making them more energy efficient.

Phosphorescent and fluorescent objects are "excited" by electromagnetic radiation and emit visible light. Fluorescence involves the object emitting light immediately after absorbing UV light. Some fluorescent materials (such as DayGlo paint, additives to laundry detergents, and some specialty papers) do fluoresce with near-ultraviolet light (wavelengths of light very close to the visible range), giving them the semblance of glowing even without use of a UV light.

When the light source is removed from fluorescent materials, they immediately stop glowing. With phosphorescence, the object typically absorbs energy within and above the visible range, "stores the energy," and releases it over a period of time even after the light source is removed.

The emission of light by matter can be explained by the behavior of electrons in the atom. Under normal conditions, the electrons in an atom reside in the lowest possible energy levels, called the ground state. If an atom absorbs energy from an external source, the energy can cause electrons to jump to a higher energy level, called an excited state. Because excited states are unstable, they typically are very short-lived. As the excited electrons return to a lower energy level, they give up their excess energy, usually as heat, but sometimes in the form of visible light.

Substances that fluoresce or phosphoresce possess electrons that are easily excited to higher energy levels upon exposure to electromagnetic radiation. As described previously, fluorescent materials emit light immediately after absorbing light of the necessary energy, while phosphorescent materials emit light over time. These two types of material differ in the mechanisms that their electrons follow to return to the ground state. In fluorescence, the excited electrons go back to the ground state almost instantly. In phosphorescence, the excited electrons drop to a lower (stabilized) but still excited intermediate level before eventually returning to the ground state. The excited electrons have a longer lifetime within the intermediate level, so this multistep process results in the observed emission of light even after the original light source has been turned off.

One commercial application of fluorescence is the addition of optical brighteners to detergents to make clothes look "whiter than white." These brighteners absorb higher energy light and then emit lower energy light in the visible range of the spectrum, making the clothes look "whiter," which is interpreted by the consumer as "cleaner." But, in fact, the clothes only appear whiter because of fluorescence.

In Part C, students make an activated zinc sulfide/glue mixture that glows in the dark due to the phosphorescent nature of activated zinc sulfide. Zinc sulfide does not become phosphorescent until it has been activated. The activated zinc sulfide you used in this activity was most likely activated by heating it to high temperatures (1000°C or 1,832°F) along with small amounts of copper or other "activators" like silver or gallium. The activators are required for phosphorescent efficiency. Copper-activated zinc sulfide glows for a longer time after exposure to light than zinc sulfide activated by other elements, which makes copper-activated zinc sulfide ideal for use in radar screens, glow-in-the-dark toys, and paints. Activated strontium aluminate is another phosphorescent material found in many glow-in-the dark products.

In Part D, students see that different colors of light have different effects on the phosphorescent material. Different colors of light have different wavelengths and different energy. Longer wavelengths of light have lower frequencies and lower energy. As the wavelengths become shorter, frequencies become higher, resulting in higher energy. White light contains all frequencies of visible light, and therefore a complete range of energy within the visible spectrum.

In the demonstration, students see that the light emitted by the red LED (a longer wavelength/lower energy color) does not have sufficient energy to excite the electrons in the phosphorescent material and cause it to glow. With the green LEDs, results may vary due to perception (some students may report the typically subtle glow generated by green LEDs as "none" while others may report it as "low") or actual differences in the glow produced. The highest-energy visible light tested is the blue LED. It creates a brighter and longer-lived phosphorescent glow.

Answers to Student Questions

Part C

Step 3

The activated zinc sulfide powder glows after the light source is turned off. The glue does not. The zinc sulfide is phosphorescent.

Reference

Sarquis, J.; Sarquis, M.; Williams, J. *Teaching Chemistry with TOYS;* Terrific Science Press: Middletown, OH, 1995.

Rubber Band Equilibria

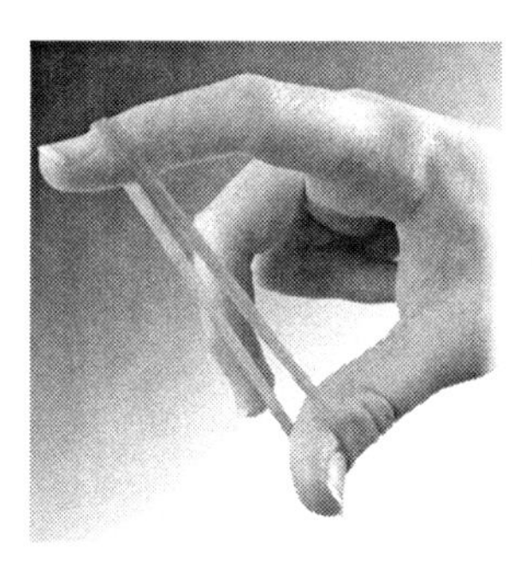

Overview

This lesson uses the elasticity of a rubber band to explore the concepts of equilibrium and reversible changes.

Key Concepts

- equilibrium
- LeChatelier's principle
- polymers
- reversible processes

National Science Education Standards

Science as Inquiry

Abilities Necessary to Do Scientific Inquiry

- *Students use logic and evidence to formulate explanations about the change in energy when a rubber band stretches or contracts. (5–8, 9–12)*

Physical Science

Transfer of Energy

- *Students observe that when a rubber band is stretched, heat is released and flows from the rubber band to the environment, causing an increase in temperature. They also observe a decrease in temperature as the stretched rubber band is allowed to contract. They learn that this is a type of energy transfer. (5–8)*

Structure and Properties of Matter

- *Students learn that rubber bands are composed of numerous polymer chains. The molecules in a relaxed rubber band have significant freedom of movement. When stretched, the molecules have less freedom of movement. (9–12)*

Chemical Reactions

- *Students use the stretching and relaxing of a rubber band as a physical analogy to the reversible process in chemical reactions. LeChatelier's principle can be applied to predict how adding or removing heat will affect a rubber band. (9–12)*

Conservation of Energy and the Increase in Disorder

- *Students observe that stretching a rubber band causes it to become warmer due to a decrease in vibrational energies; relaxing the rubber band causes it to become cooler due to an increase in vibrational energies. The heat released when the rubber band stretches is regained when it contracts. (9–12)*

Part A: Teacher Demonstration

How does temperature affect the length of a rubber band?

Materials

- 100-g weight or equivalent
- rubber band cut into a strip (A strip about 150 mm long × 5 mm wide works well with a 100-g weight. Use a larger weight with a thicker band.)
- ring stand
- heat gun or electric hair dryer

Procedure

1. Tie the rubber band strip to a ring clamp attached to a ring stand. Hang a 100-g weight from the rubber band so that the weight just barely touches the base of the ring stand or the top of the table. Challenge students to predict what will happen when the rubber band is heated.

2. Heat the entire rubber band by moving a heat gun or hair dryer up and down the length of the rubber band, as shown in Figure 1.

➤ *If the rubber band is heated too much in one spot, it may melt or break.*

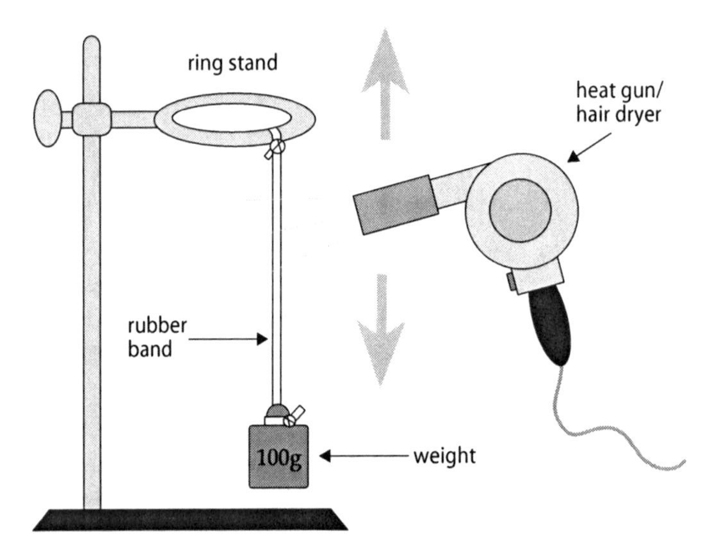

Figure 1: Heating the rubber band with a heat gun

3. Be sure that students observe the position of the weight as the rubber band is heated.

Part B: Student Exploration

Is there a change in energy when a rubber band stretches or contracts? Try the following activity and find out.

Materials

- wide rubber band
- plastic six-pack ring cut into 6 smaller pieces

Procedure

1. Examine the rubber band. When you pull on opposite sides it stretches; when you release one end it contracts. *Is the stretching and relaxing of the rubber band a reversible process or an irreversible process?*
2. Hold the rubber band against your upper lip or forehead for several seconds and notice how it feels. Move it away from your lip, then stretch it and quickly touch it to your lip or forehead again. *What change, if any, do you feel?* Hold the stretched rubber band against your upper lip. Move it away, allow it to contract, and touch it to your upper lip or forehead again. *What change, if any, do you feel? Make a claim regarding the behavior of a rubber band. Be prepared to provide evidence to substantiate your claim in a class discussion.*
3. Rubber bands are made of long polymer chains. Six-pack rings are made from polymers too. Predict what would happen if you quickly stretched the six-pack ring. Test to see if your prediction is right. Hold one of the six-pack rings against your upper lip or forehead. Move it away, then quickly stretch it and touch it to your lip again.

Instructor Notes

Tips and Instructional Strategies

- Help students write an equation for what they saw in the demonstration:

  stretched rubber band + heat $\rightleftharpoons$ *unstretched rubber band*

- Students can quantify the behavior of the heated rubber band. Use a meterstick to measure the amount the rubber band contracts. Measure the height of the weight from the table top before and after heating.
- For more advanced classes, compare the behavior of the rubber band when heated to that of metals when heated. (Heated metals expand.)
- Challenge students to relate the stretching and relaxing of a rubber band to equilibrium concepts. LeChatelier's principle states that when a stress is applied to a system at equilibrium, the equilibrium position shifts to relieve the stress.
- You may want to have students warm a 2-L, polyethylene terephthalate (PET) soft-drink bottle with a hair dryer. Ask them how the behavior of the bottle compares to the stretched rubber band.

Explanation

Rubber consists of intertwined polymer chains. When a relaxed rubber band is stretched, it transfers heat to the surroundings (including your skin in this lesson). Since the heat transferred is a relatively small amount, it is easiest to detect on especially sensitive skin, such as the upper lip. After observing this phenomenon, the reverse can be tried; a stretched rubber band is allowed to contract to the relaxed state. This time, the rubber band feels cooler, which is consistent with the rubber band absorbing energy from the skin.

It seems counterintuitive that a rubber band will contract when it is heated since most materials expand when heated. Nevertheless, contraction is exactly what happens with the rubber band. Both transformations are consistent with LeChatelier's Principle, which states that when a stress is applied to a system at equilibrium, the equilibrium position shifts to relieve the stress.

stretched rubber band + heat $\rightleftharpoons$ *unstretched rubber band*

If heat is added, according to LeChatelier's Principle, the rubber band should move from the stretched configuration to the relaxed configuration because that direction counteracts the stress and absorbs heat. Likewise, going from relaxed to stretched should liberate heat as observed.

These observations are consistent with the atoms in the relaxed rubber band having significant freedom of movement or vibration. (See Figure 2a.) When the polymer is stretched, its atoms have less freedom of movement and, therefore, less vibrational energy. (See Figure 2b.) The rubber band loses heat when it stretches and gains heat when it contracts. Stretching the rubber band causes a decrease in the vibrational energies of the molecules making up the rubber band. Because energy is conserved, the lost vibrational energy is converted into heat, which is evidenced by a temperature increase. Conversely, relaxing a stretched rubber band allows the rubber band to contract, which results in the conversion of heat from the environment back into vibrational energy. This conversion is evidenced by a temperature decrease.

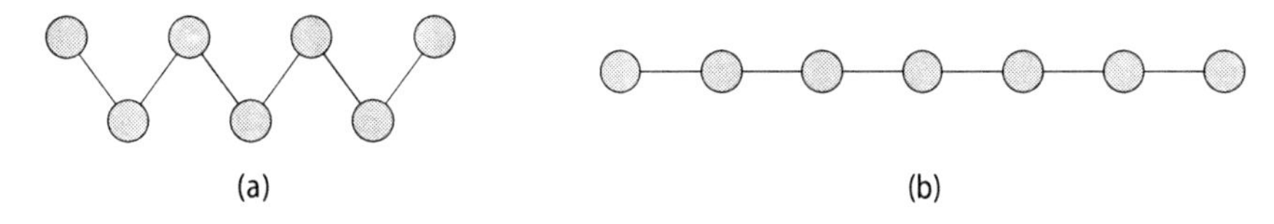

Figure 2: Polymer unit when (a) relaxed and (b) stretched

Answers to Student Questions

Part B

Step 1

Stretching a rubber band is usually a reversible process. (Breaking the rubber band is an irreversible process.)

Step 2

a. *The rubber band feels warmer after it is stretched.*

b. *The rubber band feels cooler after it contracts.*

Activities Indexed by National Science Education Standards: Grades 5–8

	A Cellophane Toy and Its Wrapper, pg. 13	Where Did the Water Go?, pg. 25	Syringe Investigations, pg. 35	Toys Under Pressure, pg. 43	Sticky Balloons, pg. 49	Degassing Soda Pop, pg. 53	Creaking Plastic Bottles, pg. 61	Investigations with Hand Boilers, pg. 67	Boiling in a Syringe, pg. 73	That Cold Sinking Feeling, pg. 79	Visualizing Matter, pg. 83	Modeling the Behavior of Water, pg. 93	Magic Sand, pg. 101	Colorful Lather Printing, pg. 109	Making and Breaking Emulsions, pg. 117	Curdled Milk, pg. 129	Investigations with Reusable Heat Packs, pg. 135	Dissolving Energy, pg. 145	Pencil Electrolysis, pg. 155	Investigating a Self-Inflating Balloon, pg. 163	Alka-Seltzer Poppers, pg. 175	Penny Sandwiches, pg. 183	Things That Glow, pg. 187	Rubber Band Equilibria, pg. 199
Science as Inquiry—Abilities Necessary to Do Scientific Inquiry																								
Identify questions that can be answered through scientific investigations.	•	•	•	•				•	•			•	•							•				
Design and conduct a scientific investigation.	•	•	•		•	•	•		•				•			•	•	•		•	•		•	
Use appropriate tools and techniques to gather, analyze, and interpret data.	•		•	•		•										•	•	•			•			
Develop descriptions, explanations, predictions, and models using evidence.	•	•		•		•	•			•	•	•		•	•	•		•	•	•	•	•	•	•
Think critically and logically to make the relationships between evidence and explanations.					•		•			•	•	•		•				•	•		•	•	•	•
Communicate scientific procedures and explanations.			•				•						•			•				•			•	
Use mathematics in all aspects of scientific inquiry.							•										•	•						
Physical Science																								
Properties and changes of properties in matter	•				•	•	•	•	•	•	•		•	•	•	•		•		•	•	•	•	
Motions and forces												•									•			
Transfer of energy	•									•							•	•	•	•			•	•
Science and Technology—Abilities of Technological Design																								
Identify appropriate problems for technological design.	•																			•				
Evaluate completed technological designs or products.	•	•		•									•									•		

Activities Indexed by National Science Education Standards: Grades 9–12

	A Cellophane Toy and Its Wrapper, pg. 13	Where Did the Water Go?, pg. 25	Syringe Investigations, pg. 35	Toys Under Pressure, pg. 43	Sticky Balloons, pg. 49	Degassing Soda Pop, pg. 53	Creaking Plastic Bottles, pg. 61	Investigations with Hand Boilers, pg. 67	Boiling in a Syringe, pg. 73	That Cold Sinking Feeling, pg. 79	Visualizing Matter, pg. 83	Modeling the Behavior of Water, pg. 93	Magic Sand, pg. 101	Colorful Lather Printing, pg. 109	Making and Breaking Emulsions, pg. 117	Curdled Milk, pg. 129	Investigations with Reusable Heat Packs, pg. 135	Dissolving Energy, pg. 145	Pencil Electrolysis, pg. 155	Investigating a Self-Inflating Balloon, pg. 163	Alka-Seltzer Poppers, pg. 175	Penny Sandwiches, pg. 183	Things That Glow, pg. 187	Rubber Band Equilibria, pg. 199
Science as Inquiry—Abilities Necessary to Do Scientific Inquiry																								
Identify questions that can be answered through scientific investigations.				•				•	•			•	•							•				
Design and conduct a scientific investigation.	•	•	•		•	•	•		•				•		•	•	•	•		•	•		•	
Use technology and mathematics to improve investigations and communications.	•		•			•	•									•	•	•			•			
Formulate and revise scientific explanations and models using logic and evidence.	•	•		•	•	•	•			•	•	•		•	•	•		•	•	•	•	•	•	•
Recognize and analyze alternative explanations and models.								•															•	
Communicate and defend a scientific argument.			•				•						•			•				•			•	
Understandings about scientific inquiry	•	•	•		•	•									•		•	•		•	•			
Physical Science																								
Structure of atoms											•	•											•	
Structure and properties of matter	•	•	•	•	•		•	•	•	•	•	•	•	•	•		•	•	•	•				•
Chemical reactions						•					•					•		•	•	•	•	•		•
Motions and forces																					•			
Conservation of energy and the increase in disorder							•	•	•								•							•
Interactions of energy and matter	•																•		•				•	
Science and Technology—Abilities of Technological Design																								
Identify appropriate problems for technological design.	•																			•				
Evaluate the solution and its consequences.		•						•					•									•		
History and Nature of Science																								
Historical perspectives							•																	

Activities Indexed by Key Concepts

Printed in the United States
145829LV00001B/3/P

9 781883 822552